NIGERIAN TELEVISION

NIGERIAN TELEVISION
FiftyYears
of
Television in Africa

Oluyinka Esan

AMV Publishing

Published by
AMV Publishing
P.O. Box 661
Princeton NJ 08542-0661
Tel(s): 6095770905 & 7326476721 Fax: 6097164770
africarus1@comcast.net & africarus@aol.com

Nigerian Television Fifty Years of Television in Africa
Copyright © 2009 Oluyinka Esan

Book & Cover Design: AMV Origination & Design Division
Cover artwork: Dapo Ojo-Ade

Library of Congress Control Number: 2009907946

ISBN: 0-9766941-2-3 (10-Digit)
         978-09766941-2-0 (13-Digit)

Table of Contents

Tables, Maps and Charts

Acknowledgments

A project of this magnitude could only be possible by the grace of God and with the help of many. I have been really blessed on both scores and for this I am grateful. The nucleus of the work was undertaken with the grants received as a scholar of the Association of Commonwealth Universities. The organisation thus deserves credit for this output and so does Professor J. E. T. Eldridge under whose supervision the project was initially developed. I cannot begin to reel out names of colleagues and friends who supported me whilst I was in Glasgow but to each one I give thanks. I must acknowledge the University of Lagos which granted me the leave to complete the study and colleagues from the Department of Mass Communication who had to adjust in my absence.

Without the grant from the University of Winchester, the second phase of the study conducted in 2008 may have been impossible. The institution and its Research and Knowledge Transfer Committee thus deserve credit, as do colleagues like Paul Manning and Leighton Grist in the School of Media and Film Studies who recognised the merit in the project.

Not much would be achieved without access to the experiences that have been documented here. For this reason I am greatly indebted to the staff of the various NTA stations, the state government-owned and private stations involved in the study, and the National Broadcasting Commission who at various times let me into their offices sharing with me their experiences, their joys and frustrations, anxieties and fears. These fall into two categories, those who were still in employment at the time of the inquiry and those who had retired. The latter are very special as they opened up their homes, in some cases at very short notice, altering their plans and allowing me in to share their reminiscences, their vision, their achievements and challenges on the job. These veterans deserve praise for their continued interest and support of the television industry.

Worthy of mention are Chief Segun Olusola, who saw the merit in my mission and encouraged me to pursue it. The former Directors General of the NTA—Vincent Maduka, Ibrahim Mohammed, Shyngle Wigwe—who granted me interviews or pointed out those sources crucial to the study are also worth mentioning. Besides giving an overview of television in Nigeria, they shared particular experiences in the regions where they had cut their teeth. I am immensely grateful for their critical insight.

Other veterans to whom I am indebted include Boniface Ofokaja and Mrs Mariam Okagbue; without them the Eastern Nigeria story could not have been so animated. I am also grateful to Francis Ejike Owoh and others at ESBS Enugu and those who granted me audience at the NTA Enugu. With their help, the gap that had characterised the story of television in the east has been filled. Without the lead given by my teacher, Professor Onuora Nwuneli, and the guidance of another veteran broadcaster Kevin Ejiofor, without the quick thinking of Peter Achebe (NTA Enugu) and the permission of her daughter Chinyelu Anyiam Osigwe, my search for Mrs Okagbue one of the living witnesses of the station that wanted to be second to none may never have yielded any result.

Ladan Salihu deserves credit for the insight into the story of the Broadcasting Comany of Northern Nigeria as does Yunusa Aliyu of NTA Kaduna. Their accounts were complemented by that from other broadcasters from other parts of the north.

I am especially grateful to those who painstakingly explained processes and policies of the various stations. Top on this list are Jimmy F. Atte and Dr. Biodun Shotumbi, both are now retired from the NTA. Likewise, Jimi Odumosu who retired from Lagos Television and Ayodele Sulaiman from OGTV who gave insight into practises at OGTV and the challenges peculiar to a particular era of television in Nigeria. In the later study Lekan Ogunbanwo of LTV and Demola Odusanya of Gateway Television ensured that the stories could be brought up-to-date. Yiljap Abraham of PRTV Jos and his team were particularly hospitable.

The private broadcasters were particularly difficult to track down, but John and Sola Momoh and Teresa Essien were helpful, once they had been found. At AIT Abuja, Imoni Mac Amarere and Tosin Dokpesi were very helpful as were other members of AIT family.

The following individuals who belong to the NTA family deserve special mention—Taiye Ogundiran in Ibadan, Biodun Balogun and Ladi Lah in Abuja; Ibrahim Damisa in Jos, with Stanley Mshelia and Peter Ochigbo; Umar Abubakar Dembo, Maimuno Garba and Saudat Suleiman in Kano, and of course the team in Zaria. The telephone interviews with Sadiq Daba and Peter Igho though brief were insightful. These and many others who cannot (or would not want to be) recognised all made the research effort worthwhile.

There are a number of friends to whom I owe a ton of thanks just for being there or for casting their critical eyes on the manuscript and suggesting ideas on how to improve the story. I am indebted to Maggie Andrews and Omawumi Efueye in particular and also to Emma and Julian Stuart. Bolade Oyebolu and Ganiyu G Alabi have helped to champion this cause especially among the Government College Ibadan old boys circuit. They have also invested in the production of the book as has AMV Publishers with Damola Ifaturoti offering his understanding and support. To these I am truly thankful.

My appreciation to my family and friends knows no bounds. I dare not make a list lest the acknowledgement becomes longer than the book. (If you look closely enough each one of you will see your names etched in gold). All my sisters and brothers made adjustments to help me get through the work in one way or another. My big sister, Temitope Adeniji Adele must be singled out for appreciation; she has been looking after me since I started school and has not stopped even now. I am truly blessed with those who prayed and worked to see me succeed. Many thanks to you all, and to my wonderful children—Ayomide my in-house editor and critic; Olurotimi my technical adviser and Olufolabomi my cheerleader who has patiently endured my gruelling regime of writing, being convinced that I will become famous when the work is done!

Preface

With revolutionary trends witnessed in media technologies since the twilight of the 20th century, it is necessary to record for posterity, some of those features of living which are subject to change. They marked an era that may soon be forgotten. Today many, especially the youth, are armed with an array of media and communication technologies—terrestrial and satellite television, SMS, twitter, email, even social networking and it may be difficult to imagine a life without these Yet the earlier technologies were forerunners of those which currently mesmerize the world; for that reason the study of television is still relevant in this time. If the curriculum in media or mass communication studies is overly concerned with new media with a diminishing interest in the predecessors of these, society is likely to lose sight of the critical paths through which the media have passed and the vital lessons that such experiences bestow. That would be a great disservice especially for those who have to manage the media today and those who rely on it for their work in society. They should not have to reinvent the wheel, rather they can learn from the experiences of others.

My fascination with television began as a child in Nigeria. I was privileged to be a part of the television generation and was so impressed by the medium that it provoked me to pursue an academic career in Mass Communication. The decision to study such a subject as the media may appear ridiculous in certain quarters, as some wonder what there is to study, and consider there to be more serious subjects to explore. The media, particularly television, appears to be primarily for entertainment. How could investment in studying the media compare with education, the economy and law, ventures seemingly much more profitable for society? What do the media have to contribute? Embedded in these queries is a basic misunderstanding of the far reaching scope and influence of the media. A critical section of society, that is, those in government, appear to have recognised this potential as shall be revealed in this book. By documenting the story of television we present here a basis to examine just what television is, with ample evidence of how ubiquitous the medium is.

Collating an account of fifty years is no mean claim, neither was the research effort. The formal research was conducted in two main phases. The first phase was conducted between January and March 1991. This was part of a larger study about audiences in a Nigerian context. The main concern then was to identify what women were watching and how they make meaning of the messages from television. Titled *Receiving Television Messages* the complete study is a University of Glasgow doctoral thesis. Due to the desire to generate rich qualitative information about the processes, experiences and explanations for the structures that evolved, ethnographic research techniques were employed in the study. That study was limited to what has been described here as the *first and second waves* of television in Nigeria, and it focussed primarily on the stations serving a particular area of the country. Because of its peculiar location, Sagamu was in the close proximity of several television stations in the South West of Nigeria—Lagos, Ibadan and Abeokuta. This (the South West) incidentally is the hub of television industry and it helped to establish the trends in the industry. Existing documentation about television in Nigeria tend to be restricted to the discussion in this area, especially the premier station (Western Nigeria Television-WNTV) and also the National Television Authority-NTA. Information from such publications was augmented by the interviews and observations conducted at the time. In that phase of the study a total of 52 scheduled interviews was conducted with a range of personnel in seven television stations and officials from allied institutions—advertising agencies and the Ministry of Information for example. There were scheduled observations of some of the key processes, for example the NTA's Network News editorial meeting, as well as the Programmes Department meeting in certain NTA stations. There were also observations and focus group discussions with core personnel in state government owned stations. In other cases, observation was incidental, occurring merely in the process of interacting with individuals within the stations. A lot was gleaned from the comments made by current and ex-staff at the selected organisations. This sort of information had to be used with caution, although such comments sometimes helped to clarify the meaning behind certain observations and more guarded official comments. At other times they were useful leads that informed probing questions during the scheduled interviews.

The second phase of the research conducted in July 2008, was designed to cover the inquiry into the experiences of television in other parts of the nation from the inception of the service. In addition the inquiry was to examine the developments in all parts of the nation since the deregulation of broadcasting in 1992, thus bringing the available record of television in Nigeria up-to-date. This period has been described as the *third phase* of Nigerian television. As with the earlier study, this was also an ethnographic study, relying largely on interviews, discussions and observations of practices. The design was intended to cover all six geo-political zones, however due to logistics constraints the study visits had to be restricted to a sample of stations in four of these. In all, nine cities, namely Lagos, Ibadan, Abeokuta, Enugu, Abuja, Kaduna, Zaria, Kano and Jos were visited. The South-South and North-East zones had to be excluded. However, since some of the interviewees had worked in those areas, there was adequate assurance that the general trends which evolved from the interviews reflected the experiences from those areas as well.

Fifty four formal interviews and discussions were conducted with personnel from fifteen television stations, veteran broadcasters, regulators and audience research bureau. This is not counting the casual discussions which as mentioned earlier prove to be most beneficial in ethnographic studies such as this. In addition the analysis of station documents, in-house publications and others, like press reports gave further insight into the practices and experiences of organisations and the broader trends in the industry. In one instance there was an observation of a press briefing by a CEO of one of the private television stations. Crucial at this stage of inquiry, was the examination of documents and data from the National Broadcasting Commission which was established by decree in 1992. This, as well as the commercial media planning data was important in appreciating the audience's perspective, as the focus of the study to this point had been limited to encoding and not reception practices. All the above research methods were supplemented with the personal experiences of interacting with the various stations as a teacher of Mass Communication in a Nigerian university and a consultant for non-governmental organisations in Nigeria especially between 1993 and 2000.

With this design, the book is poised to be a most comprehensive account of television practice in an African context, but there are two clear limitations. The first is the fact that this is essentially a Nigerian story. By delving into the minutiae of the Nigerian experience, there was no scope for comparison with other African television broadcasters. Yet the effort should be regarded as a roadmap for future inquiry into the accomplishments and challenges that confront television broadcasters on the (almost) forgotten continent. The other limitation is also as regards the scope of discussions. This inquiry has not touched on the area of Sports though it is a central service for the viewing pleasure of a section of the audience and also to the financial health of stations. This omission may be regarded as grave but it is a story rich enough to command exclusive attention that future efforts may give.

The above observations reveal the perspective, from which the account is presented, the lenses through which the story of television is refracted. This should be expected in any historical account. The accounts presented here have tried to focus on the institutions and structures as the culmination of the story in Chapter 6 reveals. Though much attention had been given to personal accounts and experiences, this was not with the view to exalt particular individuals above others who were not mentioned. Names had been mentioned to lend credence to the information. At other times, names were withheld to protect the identity of those who had spoken out with candour, to avert any possible reprisals, reduce the controversy that such identities may generate and keep the focus on scholarship. That after all is the aim of the book. A book of this nature should not seek to identify heroes or villains of history, but to tell it as it is in the knowledge that television is nothing without team work. In like manner, the book has aimed to be more than a highlight of great moments, in recognition of the importance of process in production. These considerations had shaped the structuring of the book.

A final point worth mentioning in this preface is the dilemma posed by the intended audience of this book. Because it seeks to appeal to an international market which includes those unfamiliar with the Nigerian terrain and the local politics, a deliberate attempt was made to provide the backdrop to the issues. Some readers may thus find themselves confronted with details that they may regard as obvious, however this need not be so. Such details should help informed (and not so informed) readers to see the intricate relationship between the media, politics and every day life. And when the more informed are tempted to complain that the book is stating the obvious, they should remember that some of the other readers (including a new generation of Nigerians) were not privileged to be living witnesses of much of the fifty years of television in Nigeria.

Second World War, the themes were timely and relevant. They also fitted with the concern for social and economic development championed by scholars such as Lerner (1958), Klapper (1960) and Schramm (1963) (see Chu, GC 1994; Moemeka, 1994). Like Nigeria, these nations consisted of traditional societies in transition. (Chu, GC; Schramm, A and W 1991; Reeves, G 1993) These themes pervade different theories of media and society regardless of their orientations; supporters of Modernisation theory as well as its critics, those who advocate breaking the Dependency that reliance on Western technology and such models of social advancement bred.

Fig.1 .1: Nigeria: Geopolitical Zones

Although at independence in 1960 there were three main administrative regions (North, East and West), today there are 36 states and a federal capital territory. These are organised into six geopolitical zones, suggesting that there is considerable measure of affinity within these. These zones (South-West, South-East, South-South, North-West, North-East, North-Central) reflect the original regional structure while acknowledging new power bases that have since asserted themselves and the merit for their autonomy. An example is the South-South zone. It was carved out of the oil-rich Niger Delta area that hitherto had been largely under the Eastern region, though its population is distinct from the Igbos, the dominant group in the East. The South-South itself is a diverse mix of people who, having certain common challenges, agree to work together.

From the outset, the Northern region was by far the largest region in the original structure. It reflected the hegemony of the Hausa/ Fulani ethnic groups which was already in existence at the commencement of British rule. The Hausa elite held sway over much of Northern Nigeria, establishing city states, exacting toll from other ethnic groups, imposing their language (and later their religion). The Hausa states were a formidable force but they had to align with the nomadic Fulani following the Fulani jihad and their adoption of Islam. Indeed, the British employed relics of the Hausa states, and the Fulani's Sokoto caliphate in their indirect system of administering Northern Nigeria. (Falola & Heaton 2008: 115) To the outsider, Northern Nigeria appeared to be a homogenous entity, yet this construct masked a range of ethnic, religious and other divisions. As the quest for even distribution of resources, social development, self-assertion and access to political power progresses, the fault lines in the polity become apparent.

Most Nigerians dwell in rural areas. Usually rural life is deemed to be idyllic, the pace of life tends to be slow, and people are closer to the land, making their living through a range of farming activities. As typical of traditional (rural) societies, people are largely superstitious, quite parochial and distant from the base of economic and political power. They may not seem an attractive lot for the business of television, yet their myths and landscapes make good television. Their compliance with government programmes also matters, and their votes should count. Other practical reasons may prevent them from being the targets of television. Urban life tends to be associated with bright lights and show-business, with a variety of economic activities and political issues that direct social affairs. Urban populations tend to be savvier, perhaps because they are more exposed to different cultures. They may also have the buying power that investors seek. Television seems to be made just for them. Such broad strokes do no justice to the complexity of Nigeria's population. In this nation, there are no such clear demarcations, though the rural areas are stark contrasts to the densely populated urban areas. Over the years there has been a flight of labour from the rural to urban areas yet the rural mindset remains with many urban dwellers, especially among the urban squatters.

Spatial distribution of Nigeria's population belies the concentration of opportunities for employment in industries and commerce, and the availability of communication. There are large areas of dense population in the North, and Kano remains one of the two most densely populated cities in Nigeria. In the 1950s and early 1960s Ibadan was the second most populous city in Africa (after Cairo). Though it is still densely populated, Ibadan has yielded its rank as the most densely populated city in Nigeria to Lagos, which in 2005 had a population of 9.2 million. Other major urban settlements in Nigeria include Abuja, which is now the Federal Capital Territory, Jos, Kaduna, Ilorin, Benin City, Port Harcourt, Onitsha and Enugu. This pattern of settlement reflects the types of lifestyles, time for leisure and disposable income available to people. These cities, particularly those commercial centres in the South, have an overtly exciting social scene and something for everyone – street parties, music shops, clubs, eating places and religious gatherings of different shades. Lagos, Benin and Port Harcourt, with the high presence of expatriates and their petro-dollars, set the pace, especially as regards Western influences. The Federal Capital Territory likewise; it is the newest city and has a high concentration of the wealthy and influential. The dominant Islamic influence in the North tends to be restrictive but Kaduna and Jos had been known to be more liberal in orientation except for a recent public reassertion of Islamic values and civic unrest in these parts. The social inequity prevalent in Nigeria is evident in all these cities. These are all factors that inform the business of television, as do the historical antecedents of the people.

Disparities in the historical, colonial, religious, educational and social experiences of the people are evident in contemporary social realities. Long before Nigeria was subjected to colonial rule by Britain, its various ethnic groups were subjected to different influences: ethnic origins, associations from migratory sojourns and trade and religious affiliations. Indeed, the entity known as Nigeria was formed during the colonial era, when the boundaries of the country were outlined as protectorates of Britain. Some pre-colonial inter-group linkages meant that boundaries between ethnic groups were rather fluid. According to Ajayi (2000), the British failed to appreciate such linkages. Instead they imposed new ones, especially in the amalgamation of the northern and southern protectorates, thus accentuating the North-South divide. The colonial relationship in the two protectorates differed.

> ... the British in practice tried to run Northern and Southern Nigeria as two separate countries: Northern Nigeria predominantly Muslim, with Hausa as lingua franca, and the activities of missionaries restricted; Southern Nigeria, predominantly Christian, with English as lingua franca, and activities of missionaries encouraged... Every Nigerian was classified into a tribe... While colonial anthropologists were busy reconstructing new tribal entities, colonial administrators were reconstituting pre-colonial states and polities into Native Authorities for the purposes of Indirect Rule.
> (Ajayi, 2000: 269/70)

History shows that there were glaring distinctions in the way the different groups were classified and positioned within society. People's prospects and the "privileges" they could expect were determined by the British rulers' perception of them. This was seen in recruitment in the army and the style of political administration. While traditional structures in the North were employed and strengthened, in the South the new political administrative structures weakened the traditional ones. These policies determined the strategic positioning of the elite and other classes both in the North and the South. There were disparities in access to educational or economic opportunities and actual attainment. All these were complicated by the cultural orientations to age, gender and resource-sharing. These precedents are still reflected in the competencies that emerged. For example, the NDHS report shows disparity in education and the use of media (NDHS 2003: 27-30). Whereas in the South-East and the South-West, as many as two out of every three women had attended secondary or higher school (67.5%; 65.2% respectively), fewer than one in five women in the North, particularly in the North-West (13.1%) and North-East (15.7%), had secondary education. Of all the zones in the North, North-Central had the highest percentage of women with secondary education (34.1%) yet it lagged behind the South-South, which had the lowest percentage in the South (61.7%).

The situation was marginally better for men. In all the zones, more men had secondary education – North-East (35.7%) North-West (32.2%) North-Central (63.1%) Again, the North-Central zone fared best among the northern zones but it only comes close to the South-South, where 65.6 per cent of men had secondary education. The South-East (71.4%) and the South-West (72.1%) are still ahead with regard to the percentage of men with a secondary education. These details are hidden when one looks at national averages for men (52.7%) and women (37%).

The pattern of educational attainment appears to direct media consumption in general. People with secondary and higher education are more likely to watch television, listen to radio and read newspapers. As many as eight out of 10 women with higher education watch television at least once a week, compared with one out of 10 of those with no education. This was slightly less than the number who listen to radio, but higher than those who read newspapers. On the whole, fewer men were likely to watch television, though their educational attainment still tended to affect their likelihood to watch television. A good percentage of men who had secondary education (45.1%) or higher education (57%) watched television at least once a week.

The rural/urban disparity is evident in the data in terms of educational attainment and use of the media. It is an indication of the imbalance in provision of infrastructure in the rural areas. Electricity and water supply, motorable roads, schools, telecommunications, health-care delivery facilities and the quality of residential accommodation tend to be better in urban areas than in rural areas. In any case, people tend to subscribe to both traditional and modern ways of life, irrespective of their residence. Consequently, harmful traditional practices that undermine development goals persist.

Table1.1: 2003 Level of Schooling & Literacy - Women

Background characteristics WOMEN	Secondary school or higher	Primary school or none but reads part of or entire sentences	Primary school or none who cannot read at all	Percent literate
Age				
15-19	49.8	11.5	37.7	61.3
20-24	48.3	8.1	42.4	56.4
25-29	41.3	10.6	46.9	51.9
30-34	30.4	10.9	58.3	41.3
35-39	25.7	14.2	58.9	39.8
40-44	17.5	13.6	67.8	31.1
45-49	10.0	12.4	75.8	22.4
Residence				
Urban	55.6	12.0	31.6	67.5
Rural	27.3	10.7	60.8	38.0
Region				
North-Central	34.1	9.3	55.2	43.4
North-East	15.7	9.9	72.9	25.6
North-West	13.1	7.8	78.6	20.9
South-East	67.5	18.1	14.1	85.6
South-South	61.7	13.3	22.6	75.0
South-West	65.2	13.9	20.5	79.1
Total	37.0	11.1	50.7	48.2

Table 1.2: 2003 Level of Schooling & Literacy - Men

Background characteristics MEN	Secondary school or higher	Primary school or none but reads part of or entire sentences	Primary school or none who cannot read at all	Percent literate
Age				
15- 19	64.6	14.6	18.6	79.2
20 – 24	67.9	12.4	15.0	80.3
25 – 29	59.4	19.6	16.8	78.9
30 – 34	58.3	19.4	20.7	75.6
35 – 39	45.7	24.6	23.7	70.4
40 – 44	42.3	22.3	27.3	64.6
45 – 49	31.1	28.9	30.1	60.0
50 – 54	22.6	32.6	37.9	55.2
55 – 59	14.3	32.4	42.1	47.2
Residence				
Urban	65.8	21.0	9.3	86.8
Rural	45.0	19.0	30.0	64.0
Region				
North Central	63.1	12.1	24.6	75.2
North East	35.7	24.3	37.8	59.9
North West	32.2	23.5	27.2	55.7
South East	71.4	21.5	7.0	92.9
South-South	65.6	14.9	18.1	80.5
South West	72.1	20.9	5.9	93.0
Total	52.7	19.7	22.3	72.5

2003 Level of Schooling & Literacy derived from NDHS 2005

Table 1.3: 2003 Types of Media Exposure - Women

Back ground character--stics	Reads N/paper at least once a week	Watches TV at least once a week	Listens to radio at least once a week	Uses all three Media	No exp-osure to specified Media
Age					
15- 19	12.2	41.3	58.4	10.3	34.4
20 – 24	17.7	40.9	65.0	14.7	28.9
25 – 29	12.8	39.4	61.5	10.3	30.8
30 – 34	10.6	31.2	57.7	9.5	38.3
35 – 39	9.8	33.2	59.7	7.7	37.5
40 – 44	7.6	24.8	51.7	6.1	45.0
45 – 49	6.2	23.8	51.1	5.3	45.4
Residence					
Urban	21.1	63.1	73.0	18.6	18.6
Rural	7.3	21.6	51.8	5.5	5.5
Region					
N. Central	7.7	28.0	44.7	6.4	48.6
N. East	4.8	15.9	34.1	3.0	61.0
N. West	6.7	23.8	70.9	5.6	27.0
S. East	25.9	50.9	69.4	21.4	24.2
S,South	19.2	53.9	59.2	16.1	30.9
S. West	18.5	63.5	78.3	16.5	15.6
Education					
None	0.1	12.0	47.2	0.1	50.2
Primary	3.7	32.0	54.5	2.7	39.2
Secondary	24.8	62.0	72.9	20.5	18.3
Higher	59.0	81.2	88.0	51.3	5.6
Wealth Quintile					
Lowest	1.0	3.6	33.5	0.4	64.8
Second	3.2	8.0	45.0	2.1	53.0
Middle	5.2	17.8	57.7	2.9	39.7
Fourth	13.4	52.6	70.6	10.4	19.8
Highest	33.2	86.8	83.1	30.3	6.2
Total	12.1	35.9	59.2	10.0	35.3

Types of Media Exposure - WOMEN derived from NDHS 2005

Table 1.4: 2003 Types of Media Exposure - Men

Back ground character--stics	Reads N/paper at least once a week	Watches TV at least once a week	Listens to radio at least once a week	Uses all three Media	No exposure to specified Media
Age					
15-19	10.6	39.6	61.4	6.9	32.0
20 – 24	12.8	37.7	60.9	8.9	33.8
25 – 29	14.6	35.8	57.7	12.2	37.9
30 – 34	11.7	42.3	63.1	10.5	33.0
35 – 39	6.1	29.6	51.7	5.7	45.7
40 – 44	13.3	32.9	53.4	12.2	39.8
45 – 49	13.3	37.1	56.8	9.2	37.4
50 - 54	6.0	41.4	58.3	5.2	33.1
55 – 59	10.4	30.0	51.9	8.7	42.6
Residence					
Urban	19.8	59.0	72.1	15.4	20.3
Rural	6.5	24.0	50.3	5.2	45.6
Region					
N. Central	8.4	34.8	50.0	6.6	44.2
N. East	3.1	13.5	31.8	1.8	65.6
N. West	7.7	27.5	70.4	5.9	26.8
S. East	22.3	46.6	72.5	19.3	25.0
S. South	14.3	51.4	53.6	11.4	35.1
S. West	21.5	61.7	77.5	16.7	15.8
Education					
None	4.9	16.3	48.9	4.0	48.5
Primary	9.1	32.1	54.1	6.9	39.6
Secondary	14.0	45.1	63.7	10.4	30.7
Higher	19.7	57.6	67.0	17.4	25.3
Wealth Quintile					
Lowest	2.1	11.0	38.6	1.5	59.4
Second	3.8	13.0	43.2	1.7	53.1
Middle	6.1	23.2	56.7	4.3	39.5
Fourth	14.8	50.6	64.2	11.4	26.5
Highest	25.2	72.7	80.7	21.3	12.5
Total	11.4	37.0	58.4	9.0	36.2

Types of Media Exposure - MEN derived from NDHS 2005

The cultures that are evident in the geopolitical zones had been accentuated by the geographical location of the people. From the rainforests in the South through to the grasslands, over the highlands (hills, plateaux and mountains), across the river basins to the fringes of the Sahara Desert, Nigeria does not lack for colour and culture. Nigerian societies are diverse in their language and cultures; there are as many as 374 identifiable ethnic groups. (NDHS 2003: 1) Each group takes pride in its cultural heritage and seeks to preserve this, while being entangled in a range of dynamic relationships with others. In modern times they need the media to facilitate dialogue. The groups need to clarify issues in cases of misunderstanding, explore options and arrive at consensus on matters regarding their collective destiny. The mass media have been vital in the pursuit of this but they have also been constrained by the complex social structure. How are the diverse interests satisfied?

With so many languages spoken in the nation, adopting any one language connotes the supremacy of a particular ethnic group. The choice of language for broadcast also has implications for how inclusive the service is; the more widely spoken the language is within a programme's area of transmission, the more accessible its content will be. When the population being served consists of several ethnic groups with several languages, the challenge of finding an appropriate local language for broadcasting is greater. If not properly handled, the situation fuels resentment rather than fostering harmony. For this reason, public discussions have largely been conducted in English, which draws in the elites irrespective of their mother tongue, but excludes the majority of the populace, who are largely uneducated in English. This buttresses the case against the compatibility of media representation with Habermas's notion of the public sphere (as discussed in Dahlgren, 1995), yet it is clear that representation in communication, as in democracy, cannot be avoided in contemporary social systems. According to Habermas, the public sphere is an arena that offers equitable opportunities for citizens to conduct reasoned, critical discourse without fear of reprisal. This issue of representation becomes particularly knotty when building audiences that consist of multilingual, multicultural societies. The challenge is about providing "socio-cultural interaction", which has been defined as

... encounters in which people act their roles as citizens and discuss social and political issues. It also has to do with the more fundamental construction of social reality at the intersubjective level. [about] ongoing interactions in a complex interplay with structural and historical factors. Norms, collective frames of reference, even our identities ultimately derive from socio-cultural interactions.

(Dahlgren, 1995: x)

All these call attention to how public opinions that shape development of society are formed, and support media representation that is as inclusive as possible.

To draw in more local audiences, stations had to rely on the more dominant local languages. As Ugboajah noted, ". . . in trying to bridge the gap in knowledge through communication of development issues, Nigerian mass media have resorted to an increasing use of local languages" (Ugboajah, 1980: 21). However, in this attempt, stations perpetuate the hegemonic struggles. The dominance of certain ethnic groups and their languages remains a grim reminder of pre-colonial subjugations, and many would prefer the use of their own language, or even English, to reassert themselves.

The problem of identifying an appropriate language of transmission was more pronounced when stations were centrally controlled and far from grassroots audiences. Language, after all, is merely indicative of wider concerns about the relevance of content. Television had the challenge of developing locally relevant programmes. To this end, it had to explore traditional media systems with which the audiences were better acquainted, and find ways of adopting these for the medium.

In spite of the continued use of traditional media systems, the modern mass media have become crucial in the conversations that occur in and among groups. The media help to preserve, project and promote the rich cultural heritage. They thus confer status on particular individuals or practices. On the whole, mass media help to establish norms, defend values and, paradoxically, help to challenge cultural boundaries, introducing new ways of coping (Dahlgren, 2000). There are economic as well as social implications of these media practices. So, far from simply being a neutral platform for inconsequential chatter, television, like other media, is influential. (Murdock, 2000)

Though television is the most expensive medium of mass communication, Nigerian leaders have been attracted by its potential. By their very nature, broadcast signals transcend the barrier of physical space. With difficult terrains and hardly motorable roads that lead to the hinterland, broadcasting was better suited to developmental communication initiatives. In a society with a low reading culture, broadcast media with the advantage and power of the spoken word have an edge in reaching the masses.

Radio – The Forerunner

Lagos was the first place in Nigeria to which broadcasting was introduced. The initial service was introduced in the 1920s as a wired service operated by the Posts and Telegraphs Department. Radio came to Nigeria first as part of the Empire Service (later known as the External Service, then as the World Service) of the British Broadcasting Corporation (BBC) in 1932. The Empire Service had been established to help to integrate all British colonies with the imperial authority (Briggs, 1995; Crissel, 2002). According to the BBC World Service website the aim was "to unite the English-speaking peoples of the British Empire." It was only in 1950 that the Nigerian Broadcasting Service (NBS) was properly established as a government department. This metamorphosed into the Nigerian Broadcasting Corporation (NBC) in 1957 with the passing of a Parliamentary Act in 1956. (Obazele, 1996: 144)

The wired service, known as the Radio Redistribution Service (RDS) or rediffusion, was available only in large urban areas, starting from Lagos but spreading to other regional centres. People had to subscribe; in return they got their boxes, which they rented. The practice can be likened to cable subscription, except that the decoding boxes that people got in those days were actual receiver sets.

According to Larkin, "The [rediffusion] loudspeaker . . . became an object connecting those present [by the box] to places elsewhere, inserting them in overlapping sometimes competing circuits of political identity where the urban competed with the regional, the regional with the national, and the nation with the Empire itself. . . " (Larkin, 2008: 49).

Indigenous music was promoted; local news was broadcast from regional stations. National news from Lagos, and—while the international service lasted—news from the empire planted listeners within a global network of other audiences, or to use Larkin's term, "a larger imperial circuit".

This was the prevalent mode of receiving broadcast messages till wireless radio was introduced. The British experiment with radio broadcasting had initially been subject to dispute, especially in the North. According to Larkin (2008), the aural space was tightly contested. The colonial authorities initially resisted the proposal to introduce radio diffusion services in urban centres. There were concerns about the viability of the service. Would there be enough subscribers? Would there be adequate electricity? Would the initiatives scale the hurdles of cultural custodians, and was the time right for the type of exposure that the service would bring? All these added to the dilemma regarding the cultural orientation that broadcasting would bring.

> The technical shape of radio emerged from the conflict within colonial rule between the pace and shape of cultural preservation and colonial transformation played out in the amplifiers and wires of the RDS. Radio was an infrastructure placed into service to meet the propaganda and cultural needs of a colonial regime. It was an information order, to be layered on top of older orders intended to enounce, through the sublime nature of its technology and the authoritative nature of its content, the power and the promise of modern life.
> (Larkin, 2008: 50)

When it was eventually accepted, listeners did not have the luxury of choice that exists today. Private subscription increased among locals and the expatriate community with time. In the North loudspeakers were installed in public spaces to facilitate reception for those who could not afford receiver sets. Domestic sets that were fixed to particular stations' broadcast signals were much clearer than those from the new wireless sets, which only the privileged could afford. The introduction of this technology, however, brought a new dimension to the debate about radio.

The choice between wireless and wired signals had to be weighed against the affordability of the sets for the intended listeners on the one hand, and the scope of service – whether it should be international, national, regional or local. This would depend on the identity that the authorities wanted radio to foster in its listeners. Similarly, the decision had to consider the amount of control the government was prepared to have or relinquish to achieve its mission. For instance, private investment was likely to limit government influence on the output of the stations. This had to be balanced against cost/benefit to government.

In the end, the decision was to introduce a service that promoted strong regional identities and a sense of national belonging (Larkin, 2008: 65) A national station was thus established in Lagos, and all of the regions would go on to establish theirs later. National news was to have input from the regions but radio was perceived to be the mouthpiece of whichever government controlled it, whether it was the imperialist government or Nigerian politicians.

The Nigerian Broadcasting Service (NBS) was established in Lagos with assistance from BBC London. This was after long deliberations over which technology would be better suited for the political mission set for broadcasting. Undoubtedly, when it was introduced in Nigeria, it was construed as a modernising and unifying influence. Radio was designed to pursue the Reithian vision for broadcasting, as was the case in the United Kingdom. Uche (1986: 37 – 39) demonstrates how the Nigerian Broadcasting Corporation (NBC initially Nigerian Broadcasting Service) was modelled after the BBC, which is widely regarded as the brainchild of John Reith.

Reith, who was greatly influenced by his disciplined Scottish, Calvinist Christian heritage, had brought an authoritarian discipline to shape radio broadcasting in the United Kingdom.

> Radio, in an organised social form seemed to be one significant and unprecedented means of helping to shape a more unified and egalitarian society. Through radio a sense of the nation as a knowable community might be restored by putting people in touch with the ceremonies and symbols of a corporate national life. A common culture might be established through programming.
> (Scannel & Cardiff, 1991: 13)

Reith, the first general manager of the British Broadcasting Company (before it became a corporation), had helped to establish the culture of public service. This consisted of the attempt to reach the greatest number of homes possible, delivering high-quality, not-for-profit programmes that would nurture and protect public taste. Broadcasting of this sort was to foster the sharing of national experiences, to develop a bank of collective memory and unite various segments of society. It was thus to be a force to propel civic engagement. This was the public-service vision often summarised as the tripartite mission of broadcasting—to educate, inform and entertain. Generally Nigerian broadcasters were (and still are) quick to rehash these as the purpose of broadcasting.

However, this heavily prescriptive model—justified on the basis that it was needed and was essential to tap into the modernising influence of the medium—was bound to collide with the interest of the Nigerian elite, who were on the path of cultural preservation and political independence. Besides oiling the wheel of the imperial government, it was expected to facilitate social transformation.

The technical partners at NBS offered support in engineering and programming. The first director-general, Mr T W Chalmers, had been the controller in charge of Light Entertainment at the BBC. Part of the remit of the staff seconded from London was to train local staff, and a radio training school was established in Lagos. Benefits of such local training cascaded down to television when it was introduced.

In 1957 the Nigerian Broadcasting Service became a corporation, ostensibly to prevent interference from government. It was modelled on the BBC and expected to adopt public service ideals. However, unlike the BBC, its funding could not be independent of government. Because the BBC relied on licence fees paid by the listeners, its autonomy from government was more meaningful. In contrast, the Nigerian Broadcasting Corporation was funded as a parastatal under the supervision of the Ministry of Information, thus preventing complete autonomy.

By the very location of the radio station, it was clear that the mission of broadcasting was aligned with governance and commerce. Though commercial broadcasting was not introduced until much later, all the stations were cited in strategic commercial and administrative centres. The Lagos station was responsible for national programmes that were to give a sense of national belonging. Regional stations were established in other strategic administrative hubs (Ibadan, Enugu and Kaduna). These were to help establish regional identity. Yet for many groups, especially those on the cultural and geographical fringes, their local identity was not addressed by these services. Neither the national nor regional spaces catered for their peculiar local needs. The location of the stations also had implications for the types of audiences that could be reached, and, by inference, the prospects of generating advertising revenue. Urban lifestyles and aspirations, including those of audiences in the relevant income bracket, offered opportunities for advertising messages.

Western Nigerian Broadcasting Service (WNBS) and Eastern Nigeria Broadcasting Service (ENBS) were in Ibadan and Enugu respectively. The stations were established by the regional governments in collaboration with Overseas Rediffusion Ltd. Northern Nigerian was served by Radio Kaduna, an arm of the Broadcasting Company of Northern Nigeria (BCNN). This was established by the government of the Northern Nigeria and two UK-based companies, Granada and EMI. Television operation was a complement to each of these radio services.

Some Nigerians were of the view that the broadcasting service in those days was meant to propagate the policies of the British government so that the people would comply with these (Uche, 1989; Obazele, 1996; Oso, 2004). A look at BBC records confirms this interpretation. During the Second World War, radio was clearly a propaganda channel, if only to counter fascists' propaganda. Radio had a role in the administration of the colonies. Up till 1959, as much as half of the content on the World Service had been material from the BBC Domestic service; with this, the BBC was creating imagined communities (inclusive of colonialists). Educating local communities in the English language (regarded as Britain's greatest asset) consolidated the argument that radio was undoubtedly used to cultivate a sphere of influence for Britain. (Briggs, 1995) Though radio was altogether cheaper and the more favoured medium, the visionary leaders recognised the merit in television as well. In the early days, television relied on the precedents of radio but it was distinct from its precursor.

Television (like other media forms) was regarded by the government and the governed as a potent tool of governance; one that no government can do without. It remained an ideal medium for reaching across language barriers when the written word or even spoken language is not discernible (as is the case in a multilingual society). However, it brought its own set of challenges. From the outset it had been regarded as an elitist medium. In the first instance, it was an expensive venture both for the broadcaster and the audiences. It was most likely to attract the more affluent sections of society—those who were literate and could be served by other media. Yet its features made it an attractive means of reaching less privileged viewers who might not understand the audio but who would attempt to make sense of the visuals—gestures, actions and demonstrations —in television messages. This was the paradox of television, but how was it resolved? The legislation that articulates the responsibility of various television stations in Nigeria clearly saddles them with the duty of giving adequate expression to the different parts of the nation and helping to integrate the various parts of society with the whole. How was this challenge tackled? This is the story of television in Nigeria. In the next chapter we shall examine the pioneering stations.

Notes

1. The peoples of Nigeria had encounters with various foreign trade, educational and religious missions – Arabs from the north, the Spanish, Portuguese and British in the south as far back as the 18th century. Following the scramble for and partition of Africa, it came under the sphere of the British. The influence of the British had grown steadily over the years, especially through the activities of the Royal Niger Trading Company. In 1849 the Bight of Biafra became a British protectorate; as did the Bight of Benin in 1852. Lagos became a British colony in 1861. In the north, the Niger District Protectorate was established under the United African Company (UAC) in 1885. By 1900 the protectorates of southern and northern Nigeria had been established, in effect annexing the entire area. The amalgamation of these two protectorates in 1914 was merely a formalisation of the colonial rule and the establishment of the entity called Nigeria.

Chapter 2

The Early Days

As established above, television penetrated the Nigerian market because the political elite were sold on the idea. One account traces the origin of politicians' fascination back to 1957/8 when the Nigerian constitution was agreed on in London. Nigerian politicians had heard themselves on radio and seen themselves clearly on television, and recognised that these were powerful media. They decided, therefore, that television and radio could not be left under the exclusive control of federal authorities. They agreed among themselves to put broadcasting on the concurrent list so that, just like the federal government, regional governments of Western, Eastern and Northern Nigeria could have the constitutional right to set up their own radio and television stations. This meant that each region had the power to own independent broadcasting systems.

The regions all went on to establish strong regional radio broadcasting services: Western Nigerian Broadcasting Service (WNBS) in Ibadan; Eastern Nigeria Broadcasting Service (ENBS) in Enugu and the Broadcasting Company of Northern Nigeria (BCNN)—Radio-Kaduna. With these, the federal monopoly of the airwaves had been broken, and the regions were encouraged to adopt the more novel, more prestigious medium of broadcasting. The different experiences of establishing television service in these areas will be presented in this chapter.

Also in this chapter there will be glimpses of how television service was conceived. This had been defined by the contexts, the prevalent cultural practices and the media of communication that had preceded it. Without an appreciation of these, the establishment of television will appear to have defied logic, especially as regards the question of reception. Already there was evidence of the unusual practices that were required in certain parts of the country to facilitate the reception of radio messages. Larkin records that

In 1944, engineers in Kano began to erect loudspeakers on the walls outside the emirate council office, the public library, the post office, and other prominent public places. The words and music coming from these speakers were radio broadcasts, mainly from England, which were captured by a central receiver and amplifier, relayed by wire to individual households and public loudspeakers and then discharged into urban space for anyone within earshot to hear. Radio which we tend to think of as a domestic phenomenon, began its life in Nigeria as a public technology. In its early years, the vast majority of Hausa could only listen to the radio by gathering at public loudspeakers at certain times of the day.
(Larkin 2008:48)

Now, that was the case with radio; and radio receivers were much cheaper than television sets. One question remains. Why did the governments bother to establish television at all if reception was going to be so challenging? The answer lies in their ambition for their society and the people, their political mission, their astuteness in the appraisal of what was then the newest media technology, and their appreciation of the existing cultures of viewing and message reception.

Nigerians were used to gathering to view performances on the streets during festivals. They would also gather to watch the itinerant travelling theatres. The mobile cinema unit was another feature that brought people together for the purpose of viewing. There is more evidence to show that Nigerians are well acquainted with public dissemination of media messages, whether visual or aural. Music merchants belch music out from loudspeakers for people to sample as they pass. Little wonder, then, that the idea of communal listening was the strategy employed by the colonial government for the dissemination of radio signals in Northern Nigeria as the service expanded to the area.

There was a measure of communal reception even within the domestic sphere with the extended family system and living arrangements. This can be attributed to the transporting to urban living, a rural mindset that respected communal living and sharing of assets. Neighbours maintained an open-door policy, and often, family resources such as radio and, later, television, were placed at the disposal of all. This practice was widespread across different parts of the nation, suggesting that in spite of the modern artefacts, these were still largely traditional societies. Those who introduced television knew that they could rely on this pattern of viewing to make the service worthwhile. Though it was a domestic medium, television in these contexts had to be regarded as a group / public medium as well. These are some of the fine details of a medium that had a hard and long road ahead of it.

WNTV - First in Africa

Unlike the British experience, in which regional service followed the London station (i.e. the central authority), in Nigeria, television service was introduced in 1959 by the regional government in control of Western Nigeria. This was the government of Chief Obafemi Awolowo, whose Action Group party was in opposition at the federal level. Given the cost of the venture, this was quite a controversial move. After all, no other African nation, not even Nigeria's federal government, had achieved the feat. In any case, television abroad (especially in Europe) was still trying to find its feet after the first and second world wars.[1]

Until 1936, when electronic (rather than mechanical) television service was introduced in the UK, the USA and Germany, much of the activity in the industry can be regarded as experimental. A range of management and technical models was tested in those initial efforts. How should the service be funded; what systems should be adopted for the receiver sets; what type of content—and, by inference, which audiences—were to be served? These were some of the concerns that appeared to prevail in the second decade of television. This was the era of the innovators[2], but it was a decade disrupted by the Second World War.

Television service only returned to the UK in 1946 after the war. Japanese TV was not restored till 1951, Italy's in 1954. The 1950s was the decade of expansion of television service; the decade of the early majority. Australia, Austria, Spain and Sweden were among the nations that got television in 1956; Saudi Arabia and Kuwait got theirs in 1957. New Zealand (1960), Ireland (1961) and Jamaica (1963) commenced their service after the Western Nigerian Television Service (WNTV). The 1960s was the decade for the late majority adopters of television. Although Egypt had relatively advanced plans to produce its television receiver sets and establish a station, these did not materialise till July 1960, when it first transmitted programmes. (Amin, date unknown; Berenger & Labidi, 2005) By sneaking into 1959, Awolowo's WNTV put Nigeria in among the early majority. Launched on the 31st of October, WNTV earned the accolade of being the first television service in Africa.

There were noble and laudable reasons for what could be regarded as an ambitious and revolutionary project. Television was regarded as a social service, strategic in the delivery of health, agricultural and educational services that were the cardinal programmes of the government. Reports of the official launching ceremony show that the Premier of the Western region, Chief Awolowo, pursued this project, being convinced that television was able to unlock the potentials in the region as it increased not only the pace, but the standard of education. Television was to project the people's culture, giving them a sense of pride; to present knowledge from home and abroad and useful models which would encourage people to make choices that were in the greater interest of society. In its Monday November 2nd 1959 edition, The *Daily Times*, a Nigerian national newspaper, reports Awolowo as saying, "Television will serve as teacher and entertainer and as a stimulus to us all to transform Nigeria into a modern and prosperous nation" (NTV Ibadan 1979: 6).

Clearly, Nigerian politicians had recognised the power of information and would guard jealously the channels of dissemination. As Premier of the Western region and a member of the federal opposition party, Chief Awolowo recognised the need for his region to have broadcasting stations that it controlled, having been snubbed on the federal government- owned radio station in 1956 when he had been denied the right to reply to a negative remark made about the Western region by a federal official. This would never have happened if the station had been under the regional authority.

It was fortuitous that while on a trip to London in 1955, his Minister for Information, Chief Anthony Enahoro, had become enchanted by the new medium of broadcasting, television. Thus began the love affair between television and leaders in the Western region. However, the dream would not have been realised without the

Fig 2.1: The Daily Times *front cover page of Monday November 2, 1959 reporting the launching of the Western Nigeria Television Service (WNTS), the first TV station in Africa on Saturday October 31, 1959. The newspaper's caption read: Chief Obafemi Awolowo, Premier of Western Nigeria, unveiling the plaque of the launching of the WNTS in Ibadan on Saturday. Others in the picture are Sir James Robertson, Governor-General and Chief Anthony Enahoro Minister of Home and Mid West Affairs*

Fig 2.2: Entrance to the very first TV station in Africa, (WNTV) now NTA Ibadan

appropriate regulatory framework. The changes in the constitution that placed broadcasting on the concurrent list, thus allowing regional governments to establish broadcasting stations, was what finally paved the way for the birth of the Western Nigerian Television Service – 'First in Africa'.

Table 2.1:
Timeline of Television in Selected African Countries

1959	Nigeria
1960	Egypt
1962	Congo, Morocco
1963	Gabon, Côte d'Ivoire, Kenya, Sierra Leone Sudan, Uganda
1964	Ethiopia
1965	Ghana
1966	Mauritius, Tunisia, Zaire
1967	Zambia, Malagasy, Niger
1968	Equatorial Guinea, Libya
1973	Togo
1974	Central African Republic, Tanzania, Mozambique
1975	Angola
1976	South Africa
1977	Guinea
1978	Benin, Lesotho, Swaziland
1979	Mali, Mauritania, Somalia
1981	Namibia
1984	Burundi, Cape Verde, Chad
1985	Cameroon
1991	São Tomé & Príncipe
1992	Rwanda
1995	Gambia, Guinea-Bissau
1999	Malawi
2000	Botswana*

* Refers to Botwsana Television (national television broadcaster), though Gaborone Broadcasting Corporation, a private concern, commenced transmission a few years earlier.

Being the seat of government of the Western region, Ibadan was the natural place to locate the station. With the range of cultural, educational and economic activities, Ibadan was perfect. It did not lack for talent or creativity among traditional performers and the budding artistes produced in the educational institutions. Notable among these are the Universities of Ibadan and Ife, whose scholarly attention to theatre arts, music, culture and language, especially the Yoruba language, imbued people with ideas for the new medium. Its proximity to Lagos, where local radio broadcasting (Nigerian Broadcasting Service) had begun several years earlier (1952), meant that it could also benefit from a corps of those with radio broadcasting experience. Western Region also had enviable administrative expertise within its civil service that rubbed off on the new station.

From the outset, government interest in television was clear, but the nature of its involvement was not. Television was introduced by a democratically elected government; it may thus be assumed that the service was informed by a liberal philosophy. However, less visible was the parallel form of governance to which most of the populace was subjected; traditional rule was more authoritarian. Television was a trophy and a tool for the traditional rulers as well, though structures for their participation were not as formalised in Western Nigeria as was the case in the North, where the traditional rulers (emirs) had been directly involved in colonial administration. As the military made incursions into government, another form of authoritarianism was further entrenched in the culture. The impact of this culture on the management of television becomes evident later.

There was yet another paramount interest in Nigerian television from its inception. This was the commercial interest. The Western Nigerian Radio-vision Service, the parent company for WNTV, was a partnership between the broadcasting corporation of the Western Nigerian government and the United Kingdom-based Overseas Rediffusion Ltd. This partnership seemed like an odd couple: one was firmly committed to the mission of social service, while the other was a commercial entity. They had differing conceptions of a profitable service; citizens' welfare, popular government and votes did not become the pounds, shillings and pence that were the concern of Overseas Rediffusion Ltd, so by 1961 they had parted ways. The company had similar business concerns in other nations. Commercial broadcasting was the order of the day, as was seen in the USA and the introduction of commercial television in the UK, through the Television Act of 1954 that ushered in ITV in 1955 (Petley, 2006:42; Corner, 1991: 4)). Commercial television in Britain was still bound by the strong public service ethos that had been established in the BBC.

By January 1962, the government of the Western region was the sole owner of WNTV and its sister station (WNBS, the radio service). However, the commercial interest had already been established. The Western Nigerian government had proved that television could be viable, especially if synergised with radio, the cheaper and more profitable medium of broadcasting. This was a clear victory over the sceptics who had worried about the costs of establishing television. WNTV's model was to become a standard in Nigeria's television experience.

There were three sources of revenue open to the station: advertising, sales and the government subsidy. The advertising revenue came from the more organised sector of the economy. The advertising sector had been operating in Nigeria since 1928 through the operations of the West African Publicity Company (later known as Lintas, then Lowe Lintas). This agency handled accounts for trading companies such as the United African Company (UAC) and emerging industries such as Lever Brothers Nigeria, Nigerian Breweries and West Africa Milk Company. Initial promotional efforts relied on the outdoor media (billboards and posters), press and personal selling. The broadcast media, particularly television, offered new opportunities to sell products and lifestyles that sustained the consumption of these products. By 1967, just after the outbreak of the Nigerian civil war (1967-70), WNTV / WNBS records show that there were as many as 32 agencies, 12 of which made regular bookings and accounted for up to 90% of total advertising revenue (Western Nigeria Radiovision Services Ltd, 1972).

The other stream of funding was the personal paid announcement, which had also been successful on radio. It was an acknowledgement of the value of the informal economy and more. This was an avenue for small businesses and individuals to broadcast their notices, invitations and scenes from social events. The social diary featuring obituaries, chieftaincy ceremonies, weddings and such ceremonies marking culturally significant milestones, for instance, was particularly well suited to television. It had the gaiety, emotions and drama from real people and communities. This was a source of revenue and culturally relevant content, long before reality television. On the whole, this was a social service, but it was also shrewd business sense—a case of watching the pennies.

The subsidy from government would appear to be the most secure funding for television, especially since this was to be such a strategic tool of governance. It was certainly needed, as expenditure far outstripped the advertising revenue. But dipping into the deep pocket of government came at a cost. It proved to be the station's undoing, as government officials expected the returns on their investment to be much more than the regular dividends of public service broadcasting. Rather than promote accountability as required of the media in Habermasian notion of the public sphere (see Seaton, 2003: Price, 1995), deference to political leaders was expected – nay, demanded. This was more than a reflection of the authoritarian culture, and marks the most dysfunctional use of the medium, as was the case during the turbulent days of politics in Nigeria's "wild west".

According to Yemi Farounbi, who became general manager of the station in 1977, the last general manager before WNTV was subsumed into the national television service,

> 1962-65 [was the phase] when television was completely politicised. The station was viewed not only as an essential arm and instrument of government, but also as a megaphone of the party in power. Given the circumstances of the politics of that time, it was inevitable that the image and credibility of the station got tarnished.
> (NTV Ibadan 1979: ii)

From the outset, the original arrangement at WNTV reflected a desire to share power. There were two managing directors: a Nigerian and a Briton. Along with top management and general operations, there were expatriates heading News, Programmes and Engineering departments. At this time, Nigerian staff, including engineers, had been hired even if they were junior within the organisation. By October 1961 there was only one managing director, a Nigerian (Olu Ibukun), while the station manager was an expatriate who was not involved in policy.

At the expiration of the initial contract in 1962, all expatriates, barring those in engineering, had left. The expatriates who remained were no longer under the joint arrangement with Rediffusion; they were on contract to work with WNTV and they held the most senior posts in this department. By this time as well, Nigerians had taken charge of sensitive departments such as News and Programmes. The collaboration with Rediffusion was thus very short.

It is alleged that one of the problems between the Western Nigeria government and Rediffusion International was based on the government's attempt to dictate policy and to interfere in the running of the station. This was not surprising, given the orientation to power and the reluctance of the Nigerian proprietors to relinquish control. Whereas the foreigners had been businesspeople who had their own profit motives, for the Nigerians there was a lot more at stake in the broadcasting business. Party politics in Nigeria dictated particular needs that the foreign partners may not have understood or subscribed to. It was easier for them to be more dispassionate. The prevalent mood at the time, at the dawn of independence, was also such that the government would have been keen for Nigerian staff to take over as soon as possible. Independence could not be complete if expatriates were still in control of the only television station in the country. This may also account for what then was deemed interference in the administration of the station.

In retrospect, WNTV needed more time to properly conceptualise the mission and how this would be delivered. The decision to commence service at the time appears to have been informed by a linear model of communication that focused largely on the transmission end of the process. The "hypodermic needle" model of communication that was prevalent in the 1950s failed to recognise that the effectiveness of media messages required more than beaming out messages to audiences. Effectiveness of messages also depended largely on what happened at the reception end. Enormous investment in transmission and message production was not sufficient guarantee of the success of a mission. In other words, television service ought to have taken a more holistic view of the communication process in recognition of both the encoding and decoding processes.

At least two dimensions to the reception process required further consideration. Having identified the right audience, their willingness and capability to receive the television signals was crucial to the success of the mission. The investment in television should be judged by the size of the audience. The initial cost of transmission was high, but if amortised against the audience size, the larger the audience, the lower the real cost. For this to happen, people needed access to affordable television sets so that the message could be received. A ground-breaking research effort initiated by the station in 1962, focused on rural audiences in particular, had shown that these audiences were enthusiastic about local programmes, even though they could not afford television sets. The Omi Adio-Lalupon TV Viewing Experiment led by Marvin Rimmerman of Syracuse University and Segun Olusola, the first indigenous television producer in Nigeria, had identified how central the establishment of community viewing centres was to the success of television's mission. This was enough rationale for setting up more community viewing centres in places such as Mapo, Omi Adio, (both in Ibadan) Ejigbo, Epe, Ijebu Ode and even more remote villages. At these venues television sets were procured for the community, powered by petrol generators and located in central places where villagers could assemble to watch. This was a way of ensuring that TV reached the audiences, but even this strategy had its limitations. It was a costly venture for the programme-makers and the viewers, who had to make the extra effort to visit the viewing centres. Television in this mode was not a domestic medium that facilitated spontaneity in viewing. It was only a matter of time before the community viewing initiatives were discontinued in the face of more pressing needs.

For audiences to engage with the message, they needed to be properly disposed to the viewing experience. This meant that the occasional exposure to a communal set was not adequate, yet this was the situation at the start of service. There was little that WNTV could do to improve that situation. Affordable sets were not available elsewhere in the world at the time, since even in the more affluent societies, television was still a novelty. There was no guarantee that deliberate government action to promote local manufacture of receiver sets (as attempted in Egypt) could flood the market with affordable sets; yet that was the only way to improve the intensity of access to the medium.

The other dimension to the issue of access in those early days might have been more manageable if the government had been willing to bide its time on the television project. Specialists were required to package the programmes properly, especially the type needed to fulfil the defined primary mission of teaching through television. This required training of the staff in the appreciation of audiences and the design of the programmes. Messages needed to be packaged in a way that would be appealing; if learning is to occur, viewers must appreciate and comprehend the messages. All these occurred later in the history of television but in the early days, much was compromised. It may have fared less well in its educational mission, and perhaps the good intentions collided with political and commercial imperatives, but one thing is clear, WNTV made a mark. It proved that it was possible to establish a service which at the time seemed like the preserve of the more affluent nations. There was a "feel-good" factor in this achievement, and the nation went on to build on this. It is ironic that television was prevented from being a democratic medium, considering the laudable justifications for establishing it in the first place. The tight regulation to which the medium was subjected lends credence to the claims of those who viewed the noble cause with cynicism. It was common knowledge in Nigeria at that time that the commendable reasons for the television project were tainted by the political motives.

The burden of technology

One of the burdens Nigerian television has had to bear since the early days is the dependency on imported technology. In spite of the disparity in the economic capacity of the nation compared to other nations involved in the television business at the time, Nigeria did not enjoy any differential charges when purchasing equipments. If anything, because manufacturers were businesses, they had every reason and inclination to charge more in shipping and handling costs. Their rationale was the prohibitive cost of trading with some *small* African market in the days when travel and communication were much more cumbersome and rather circuitous. At the time, transatlantic travel tended to be by sea, as air travel was not widely available. Even telephony was nowhere close to being straightforward, in contrast to the instant and direct global connection that satellite mobile telephony affords today. For such reasons, the cost of running businesses was higher in those days.

At inception WNTV was the only station for a radius of 4,800-6,400km (3,000-4,000 miles). Sending one component to Ibadan was not cost-effective for the manufacturers. It was a lot of trouble, as there were no local suppliers; if perchance there were, it would not be wise for them to hold stock, since the market was virtually non-existent. WNTV launched its operations before some other more technologically advanced nations. It was fishing in the big boys' pond and had to pay like the big players for equipment and spares. Otherwise, WNTV (and, later, the other Nigerian stations) had to rely on cast-offs from foreign stations that were able to update their equipment as technology developed. Such acquisitions of obsolete or near- obsolete equipment put a strain on operations, especially in the area of engineering. When the equipment wore out, which it soon did, there was no technical support from the manufacturers, which usually had discontinued the models. This was a challenge for the engineering staff, who, like the programming operatives, had to make the medium adaptable to the context of operations. In responding to such challenges, staff at the station demonstrated a measure of resourcefulness that is part of their enviable legacy to the nation, as the following case-study will show.

The case of AMPEX 2-inch videotape recorder

One example that illustrates WNTV's resourcefulness is the AMPEX quad videotape recorder, a novel technology at the time (Abramson, 2003:60 -76). It had been taken on exhibition, touring the world, and ended up at a broadcast exhibition in Ghana in 1964. It was offered for sale as partly used demonstration machine to save the costs of shipping it back to England at the end of the tour. It came with kit and spares. WNTV bought this machine at a discount and it was the only one of its kind on the entire west coast of Africa. The tragedy was that the machine did not work when attempts were made by the expatriate engineering chiefs to set it up, despite the manuals and spares that were available. The machine thus languished in a specially designated area that was fully air-conditioned to keep it at the right temperature, no more than a display piece. Plans were made to bring in representatives of the manufacturers from abroad, who would more likely have better success in identifying and rectifying the fault. Because it was a novel piece of technology, this seemed to be the logical thing to do, but the budget for the year did not accommodate this plan, so the machine remained a showpiece. According to Vincent Maduka, "It was put in a specially air-conditioned room locked and [marked] No Entry." It would have been so till the following year but for the initiative of one of the young Nigerian graduate engineers.

Driven by his curiosity and confidence in his education, he sought permission to have a go at repairing this machine, even though, as the expatriate assistant chief engineer had told him earlier, he had never seen one before. The chief engineer, who by then was a Nigerian, was willing to indulge the graduate in his curiosity, on condition that he was supervised by the expatriate assistant chief engineer. The young engineer was Maduka, who was eventually to become general manager at WNTV and later the first director general of the Nigerian Television Authority.

> I went away and read it [the manual] and then we started working on it. In no time at all he [the assistant chief engineer] left me to it. He was basically a technician that rose through the ranks abroad. I was a university graduate using first principles, so while he was lost, I was not lost. I had never seen it before but the principle was there for me and I followed it and we delivered it. . . . Anyway, we got it fixed and had a cocktail to launch our VTR.

The incident prompted a new respect, indeed a commendation for the graduate engineer who had initially been resented by this expatriate. The resentment came from the interpretation of the job—the distinction between *brain engineers* and *hands-on engineers*. Whereas those technicians who rose through the ranks tended to be more hands-on in their role, more likely to be found climbing masts, rigging gear, shining the machines, the academic orientation of the graduate engineers (perhaps coupled with the pride in their accomplishment) meant they were more inclined to planning policy and tackling challenges that require first principles. This rivalry was evident right from the early days and created the sorts of dilemma that should be considered when staffing stations.

Achieving the Mission

In the final analysis, the official mission for television was to promote education. This tied in with the administration's policy of putting education first. It was also to be a window on the world. But television operatives such as Maduka admit that this mission was somewhat compromised. Education was not central enough in the operations. If it was to take such a prominent place, education needed to be more generously funded; yet, paradoxically, the station's pursuit of revenue worked against the educational mission. By 1961, the government had begun leaning on the station to be self-sufficient, taking a cue from the success of the American model of commercial television. By leaving management to source funds, the more commercially viable entertainment sector became more prominent in the drive to generate income.

As outlined by Williams (1990), television has different forms (News, Argument and Discussion, Education, Drama, Variety, Sports) each with its own conventions that inform how they are combined together in the flow of programming. The contrasts in form become more apparent when they are positioned against each other on the schedule. Whereas entertainment tended to have a style and content easily accessed, and thus a mass appeal, education of the sort that favoured the use of the medium for lectures and lessons with even less sophisticated production and visual demonstrations, was more restricted in the size of audience it appealed to. These programmes, stressing academic factual knowledge, were elitist. Yet educational programmes, if well conceived and packaged, relying on everyday experiences or general human knowledge, could be popular (as shown by educational programmes such as the American Children's Television Workshop's *Sesame Street*). Through such educational forms knowledge can be democratised (Fiske 1987: 266 – 9). This suggests that programmes with an educational orientation can still be commercially viable but they require the right level of investment, clear appreciation of audiences and how they learn—that is, what teaching techniques work best for them.

The emphasis at this point in Nigeria was different. The education initiative focused on the ability to receive the programmes: after all, without receiver sets, even the most carefully planned productions would be lost on the audiences. This was most likely to be the case for those with the greatest need for educational programmes; they had little, if any, access to receiver sets. The strategy to combat this was to work with schools within the coverage area. Programmes were aligned with the schools' curriculum, and some schools were supplied with one or two television sets to facilitate reception in schools, which—while an expression of the commitment to education —was not adequate, and was even undermined. The joke was that not long after they were supplied, some sets disappeared into the homes of headmasters. There were other obstacles impeding reception even when keepers did not divert the sets for their personal use. Paramount among these was power supply and the cost of running petrol generators. Television could not be adopted widely without this infrastructure.

Another problem of the educational broadcasts was how to synchronise lesson times within the school day with the television stations for the schools' broadcast. Some progress was made on this as WNTV and the Western region's Ministry of Education reached an agreement on how the service was scheduled, even though it remained an add-on to both institutions. Educational broadcasting in Nigeria was a shared responsibility between the Schools Broadcast Units (SBUs) of education ministries as content providers, and broadcast stations as the channel providers. The greater burden was with the SBUs, which had primary responsibility for making the programmes. The Ministry of Education provided the didactic elements and the personnel, including the programme producers. The stations, first radio then television, provided them with facilities for production and broadcast of the programmes. The Western Nigeria government had set up its SBU in 1957, and was involved in educational broadcast services by 1958 (De-Goshie: 1986). When it arrived, WNTV provided the facilities and airtime for this venture. In those days, transmission on WNTV was limited to the evening from six o'clock to 10 o'clock or thereabouts, so the schools' broadcast was run during the day. That the two parties with responsibility for educational broadcasting were never really fully integrated hampered the prospects of using television for education. Still, WNTV in its 20th-anniversary accounts records modest success in this venture on the basis of feedback from schools.

WNTV ran programmes in Biology, Geography and Physics—subjects that had universal standards (unlike the humanities such as Literature, History and Art) and could therefore be imported. These were the subjects required as the foundation for the region's much-needed human development. The bulk of the material in those days was imported. Nigerian input was little more than talking heads, with presenters merely introducing the programmes; this model was, however, to change with time.

In addition to the formal schools' broadcast, the station recalled in its 20th anniversary memoirs its effort to produce broader-based educative programmes for young people. This included *Careers*, a programme that featured films of people engaged in various careers. *News and You* was a comprehensive, in-depth package reviewing topical world news, aimed specifically at young people; this was intended to sharpen their understanding of current affairs. *News and You*, produced in collaboration with the Audio Visual Unit of the Ministry of Education, was deemed to be the greatest success in educational broadcasting (NTV Ibadan, 1979: 41 – 42).

Without systematic inquiry or measurable results, the impact of these efforts is hard to judge because education is such a complex mission; a more difficult mission to pursue than entertainment, which was a matter of taste and preference and whose performance can be measured in terms of size of audience attracted. Because these were pioneers, television practitioners' determination in the pursuit of the initial vision helped them defy some of these obstacles, yet it was a matter of time before this waned. In any case, entertainment was also better aligned with the commercial orientation, and WNTV will be remembered for its achievements on this front.

Most remarkable of these are the local drama programmes, which tended to have didactic qualities. Some of these were series, featuring once a week, some were epic dramas, often presented around public holidays or to mark festivities. There were regular slots for Yoruba drama productions: comedy (*Alawada; Awada*), court drama (*Kootu Asipa*), folk drama (*Bode Wasimi; Tiata Yoruba*). The station was blessed with the Yoruba travelling theatres, led by distinguished performers such as Moses Olaiya, Duro Ladipo, Kola Ogunmola, Oyin Adejobi, and the doyen of theatre, Hubert Ogunde. Considering that these had no formal training in the medium, adapting their skills to television was a remarkable achievement for them and for those who produced them. In those days, there was no field recording, visual effects were limited, yet stories were not always simple. For example, how could they depict the journey through the woods or the encounter with the spirit beings; the inevitable lines in the epic dramas that were attempted as special productions? Set construction, costumes, make-up and performance may have lacked finesse by modern standards, but the popularity of these Yoruba programmes made the Herculean task worthwhile. It also meant that the station could reduce its dependence on foreign productions that dominated the screen at the inception of service.

> WNTV/NTV Ibadan can claim that it has played a leading role in the flowering of Yoruba theatre that has been a feature of the artistic scene since the late sixties. With regard to actual productions, it has not always been easy to get the artistes to accustom themselves to acting for television. Our studio is rather small and the expansiveness of traditional acting has to be reduced for television.
> (NTV Ibadan, 1979: 44)

Having mastered the transition from stage to the small screen, WNTV veterans such as Lere Paimo (Eda Onileola) are still waxing strong in Nollywood—Nigeria's video/film industry—even in the new millennium. Many of the younger performers are scripting and producing in Nollywood; evidence that they have come a long way. Much of the Yoruba drama in the early days was ad-libbed rather than scripted. WNTV should share some of this credit for its pioneering role in this sector.

WNTV also blazed a trail with dramas in English, which became regular features. From fairly early on in its operations, *WNTV Playhouse* was introduced featuring groups such as the Nigerian Armchair Theatre and Orisun Theatre. Unlike the Yoruba groups—which were 'professional' and more established, albeit for street theatre —the English performers were amateurs. Perhaps this fact, along with their proximity to resources and support from the University of Ibadan's School of Drama, meant they were more malleable for the medium. By Western standards their attempts were more cultured, in terms of the refined performances and the resonance with elite (Western) culture. Again, this was no mean feat. Here is an account of how this was achieved from Lekan Ladele, one of the pioneer actors who later became head of drama at the station.

> It was the first attempt to bring regular home-made drama on our screen. The challenges were many, but the ground work had been well-laid by Mr. Ayo Ogunlade, Controller Programmes, and Mr. Uriel Worika the pioneer director of the Armchair Theatre. Performances were transmitted live, a situation made [possible] by unusual discipline characteristic of the cast. The stars included Tunji Oyelana, Wale Ogunyemi, Seinde Arigbede, Biodun Jeyifous, Bisi Komolafe, Yewande Johnson and Jibola Dedenuola. The plays were popular and the William Tell Overture signature tune heralding each performance signified the beginning of the best in local programme productions. The scripts came from members of the group . . . A lot of enthusiasm and dedication went into the productions and it was easy to get the half hour play ready in 4 days – all lines in the head! The actors were talented, mature and ready to work extra hours. (sic) (Ladele cited in NTV Ibadan, 1979: 44)

Though WNTV wanted television ostensibly for education, it was not long before the economics of the business led to a redefinition of priorities. Television soon became "cinema in the house"; a medium for entertainment, although there is evidence to suggest that it was more than this.

From the beginning, its news and current-affairs programmes had been a "tent pole" in the schedules. In 1972, the news bulletin was watched by 99.2 per cent of the station's audiences in Western state and 96 per cent of those in the Lagos area. This popularity attests to the audience's appetite for information, as the quality of the news had been a challenge for the station from the inception of the service in 1959, when the news team was led by Peter Proudman before the exit of the expatriates. Even back then, there were Nigerian newscasters such as Anike Agbaje-Williams, Kunle Olasope, Nelson Ipaye and Yomi Onabolu. Between 1959 and 1972 the news team had no more than 15 persons, including senior editors, duty editors, sub-editors and reporters, who could hardly cover the range of issues across the breadth of the state. The station relied more on news agency reports and press releases – and the news was often hijacked by the politically appointed managing directors of the station for their party.

These experiences inspired the quest for professionalism, and during the civil war the number of local correspondents was increased, with the News Department expanded and restructured to increase and balance local coverage. By 1973 a clear set of objectives for news had been defined; more soundbites were in use, even where footage was not available; there was more creative use of graphics to complement the news and ensure that the set objectives were met. The station took the news very seriously: *Highlight, Insight, Dateline Nigeria* (later renamed *Nigeria in Action*), *The World Last Week, E da S'Oro Yi* [Yoruba programme title meaning *Join in these Deliberations*], *Issues of the Day* are a few examples of the current-affairs programmes that were featured on the station. Whether they were formatted as discussions, interviews, documentaries or in-depth reviews of the news, these were bold initiatives that helped a wide range of viewers, irrespective of their age, educational attainment or gender, to engage with topical issues.

Enlightenment programmes and jingles on WNTV / NTV Ibadan were remarkable for their attempt to address the informal education needs of rural audiences and build the information gaps that existed across sectors and generations. *Kaaro Oojire* was a vehicle for the transmission of Yoruba culture. The station pioneered the Yoruba- language ombudsman programme *Agborandun* that helped to resolve disputes and inform people about their rights and responsibilities. This format has been replicated across stations that later evolved in South-West Nigeria. *Ere Agbe* may be regarded as on-air extension farming service. By the time this programme was introduced, portable TV cameras were available and footage from farms could be used. Likewise *Ayelu*, which focused on the development and self-help projects in the rural areas, benefited from the use of ENG (electronic news gathering) cameras. There were also health programmes in English and Yoruba, and of course the Yoruba jingles that focused on health and safety issues but which must be regarded as classics in their own right.

There were programmes dedicated to women and children, although in October 1962 there was no locally produced programme for children on the schedule. However, there was no shortage of foreign programmes, .including the British *Adventures of Robin Hood* (1955-59), *Adventures of Superman* (1952-58), the American series *Lassie* (1954) and the first of the Gerry Anderson puppet adventure series, *The Supercar* (1959). There were others, such as the British classic *Adventures of Noddy* (1955), the Australian series *The Terrible Ten* (1960), ITV's sitcom *The Adventures of Aggie* (1956 -57) and more. In later years programmes including classic American cartoons like HannaBarbera's *The Flintstones* (1960), *Huckleberry Hound Show* (1958), *The Yogi Bear Show* (1961); Anderson's puppet adventure series *Stingray* (1964-65); the Australian children adventure series *Skippy The Bush Kangaroo* (1966) and many more were featured.

The situation as regards local productions for children had improved by October 1971, when *Junior TV Club* introduced by an expatriate, Mrs McGrath, featured twice a week. Elsie Olusola introduced *Bedside Stories*. Station records show that Junior TV Workshop, *Children's Variety Hour, Our Heritage, Young Voices* were examples of efforts that sought to stimulate and hone the artistic talents of children. The variety programmes had segments including story-telling, poetry recitals, quizzes, games, musical performances, drama and sometimes craft-making. This has become an established format for children's variety productions nationally; it has also been adopted for productions in Yoruba and other local languages. The challenge in these programmes was to attract and sustain the attention of the children. The foreign productions that children had become accustomed to were hard acts to follow, yet local productions also had their charm. Much of this depended on the creative articulation of ideas, thoughtful planning and hard work invested in coordinating the productions. This included scouting for loveable participants— children and other facilitators. The programmes offered opportunities for studio participation that was a treat for those children who were involved. Programmes also relied on the charisma of the presenters;

their command of language, the good manners, good humour and grace with which they made the productions appear effortless.

Anike Agbaje-Williams was one of the veterans of WNTV who gave more than glamour to the audience. She was responsible for *It's a Woman's World*, a magazine programme that sought to broaden the horizon of women in the roles that had been assigned to them in modern society. The programme featured issues reflecting traditional gender roles for women—fashion, family care and maintaining relationships within family life, home economics—but it also featured issues pertaining to health and education that would enable women improve themselves (and their children) and better contribute to the development of society. The programme was designed to be informative and entertaining. To do this, novel talents who could inspire the women had to be found on a weekly basis, as each episode had to be fresh and engaging. There were some unusual finds such as the 70-year-old *Olori* (wife of a Yoruba traditional ruler) in the *Alaafin* of Oyo's courts. This woman, due to her longevity, had spanned the reign of nine Alaafins! This was a very interesting subject —how had she served them all? In that instance, the programme would have facilitated engagement with cultural practices in addition to the usual issues it addressed. But with the achievement of regular weekly productions that kept the audience engrossed came challenges from what should have been unlikely quarters, as the producer recounted. "Difficulties arise when a producer finds himself at the mercy of the ancillary services arm of programme production. I recall an occasion when a live show was totally ruined because I could not have a camera rehearsal" (Agbaje-Williams quoted in NTV Ibadan, 1979: 48). This may have been due to the pressure of various demands on available facilities, yet it may also reveal attitudes towards the women in television. It is no surprise that programmes for women and children were the exclusive responsibility of women.

Women also featured on the entertainment programmes as presenters of the music and variety programmes *Music Extravaganza, Youth Rendezvous,* (later *Weekend Variety*) which featured an array of acts. These provided a boost for the local music and entertainment industry that supported some of the programmes such as *TYC Show*. These were shows designed for family viewing, as were *Young Voices* and *Quiz Time* (later *Take a Pick*). This may have been modelled on an ITV production of the same title. In all these, WNTV/NTV Ibadan took strides towards making television be the teacher and entertainer, the stimulus for social transformation that it was conceived as. With the expertise of the intellectual community, including the National Institute for Social and Economic Research (NISER), and feedback from its viewing public, the station regularly reviewed its operations. Pioneering staff experimented with formats and organisational structures that worked best for their audiences once the initial distraction from the political upheavals had been overcome. By the time the Nigerian Television Authority (NTA) was established, WNTV had a lot to bequeath to the nation.

Second to None—The Early Days in Eastern Nigeria

Eastern Nigeria Broadcasting Corporation Television (ENBC-TV) was established in a competitive political environment. In the first republic, there was healthy rivalry between the regions; no region wanted to be outdone. Though it came after another, ENBC-TV took the slogan "Second to None". Its operations commenced on the 1st of October 1960 as part of the celebrations for the nation's independence. The station was regarded as the Premier's independence gift to the people. The Premier of the Eastern region at the time was Dr Michael I Okpara, whose government, it is claimed, mirrored Chief Obafemi Awolowo's in many ways. For instance, both led regional governments which introduced television services that collaborated with the same overseas partner—Overseas Rediffusion Ltd. Both stations were regarded as being dynamic and having people-centred programmes. This to some extent reflects the development needs of that era.

From all indications, the Eastern initiative was quite modest. In any case the Eastern region had the smallest population to cater for compared with WNTV and its Yoruba- speaking audiences or Radio Television Kaduna (the northern regional television service), which would serve the diverse Hausa-speaking groups. (Ajayi, 2000: 264/5). Though they were fewer, serving Igbo communities was no less challenging, as they had no centralised social structure. The Igbos are reputed for their republican orientation.

At the core of the pioneering staff were three expatriates hired by the regional government. They were British, and one of their local colleagues recalls that they were Jewish. What is more pertinent is the fact that they were adaptable and resourceful and did not come with a full complement of staff. Peter Proudman, who had been in Ibadan, had overall responsibility for programming, while a certain Mr Smith was the technical officer responsible for engineering operations. Mr Humphrey was in charge of general management (administration and finance). They were described as being very enterprising, but also motivated by profit.

The first priority for this three-man team was to recruit local staff. They began headhunting within the Enugu vicinity, but also looked further afield, to Lagos and Ibadan. They took inventory of human resources available in their area of operations and considered how they would make such talent work for the new medium. They required staff with technical competencies for the medium, so they went to *borrow* from Ibadan. Staff in Ibadan (the earlier station) had been trained in the essentials of television operations. Three camera people and a director were hired from there. Of these, one of them (Bibilari) was a Yoruba man. The other two camera people were Benin (Edo) men such as Jacob Osodi, the director. They had all been living in Ibadan and had received their training there before being recruited for the Enugu station.

Another phase in the recruitment strategy was to recruit the creative staff (producer/presenters). Enugu was a civil-service city and regional centre. It had its own reserve of people who had relevant or allied experience. Some of the residents had good education, spoke well and were great ideas people. Some of them had even travelled abroad and were somewhat familiar with the medium. There were those who had speech and voice training, for example, and some had worked in broadcasting overseas or with the federal radio station (Nigerian Broadcasting Corporation – NBC). Mary Umolu was an African-American married to a Nigerian and had apparently had some experience in schools broadcasting in America. Mariam Okagbue was another who had done speech and voice training abroad and had done part-time reporting with the BBC at Bush House in London. Given the low level of black representation at the BBC at that time, this was a remarkable experience. Okagbue was experienced in making programmes about women,

Africa and Nigeria in particular. Securing such a talent was thus an advantage. She recalls being a few months pregnant when she was recruited—evidence that the recruiters had a long-term perspective of their mission. They also hired announcers who had appropriate language skills from the local NBC radio stations. Presentation was critical in those days, as most programmes were live shows, since there were no facilities for recording. If they did not have a mastery of the subject or fluency in the language of presentation, presenters were bound to stumble on air, so recruitment of Igbo-speaking staff was essential for smoothness in operations.

Like WNTV, the station also relied on the university community. It recruited from the new crop of theatre arts graduates from Ibadan. People like John Ekwerre and the late Effiong Etuk who were involved in drama productions were hired from Ibadan, an institution that was prolific in its productions of various dramas, ranging from Shakespeare to JP Clark. Occasionally they toured the country with their performances, and their talent was showcased around the nation. Proudman was a good scout for such talent. He hired staff for the different roles (presenters, producers, newscasters, and later directors) that were required for television.

The station relied on experts from the neighbouring University of Nigeria in Nsukka (UNN) and other local industries. The coal-mining activity in Enugu city and the presence of railway installations had already attracted people with a range of skills from around the country. These account for the presence of local families who were not Igbos in Enugu; "railway families", who were well travelled, at least within Nigeria. These formed another stock from which the new station was able to recruit. They offered the transferable skills required for administrative tasks at least, so the station was able to fill such posts with relative ease.

The pay was not much, but it was better than a teacher's salary; the sort of job that was readily available for those with education in those days. A trained teacher in Higher Elementary (who had to work to death), was paid £2,000 annually, yet some broadcasters were able to attract a salary of £500 a month. The job was challenging, but it was fulfilling. Perhaps what mattered more was the sense of duty and the glamour of working in this novel medium.

Many of the local staff recognised the opportunity of affecting lives; they were strategic within the operations of the station.

Right from the start it was clear that television broadcasting required devotion. It was like a family affair; staff needed to go beyond the call of duty to be accomplished, as will be discussed later. It also helped if they were versatile. Before the Nigerian civil war (1966-70), there was only one professional director in the station —Jacob Osodi. Technically skilled and highly devoted to the job— practically living in the studio – he was described as the "livewire" of the station.

ENBC-TV was intended to express the orientation of its political patrons and the cultures of the people whom it was meant to serve. The prevalent culture in the East was, however, slightly different: unlike other feudal Nigerian societies, Eastern Nigerian societies are *republican*. The people were at liberty to air their views and arrive at consensus rather than merely being told what to do. In this post-colonial situation, the desire was to make television service as widely available as possible. Transmitters were located for coverage of what was then the Eastern region, and there were studios in Enugu and Aba, a commercial centre that was also relatively close to Port Harcourt, another critical sea port in the Niger Delta. Programmes originated from Enugu and Aba were transmitted to the different ends of the Eastern region and beyond.

The substation was needed because the people of Aba insisted on having their own station, in the same way that they insisted on having their own state many years later. These were a people who were conscious of their (distinct) identity, and having a television station devoted to them was just one way of expressing this. So among other functions that it fulfilled, television was used to mark group identity, but this came at a cost. The Aba substation was run like a station, yet most of the producers and directors came from Enugu. These, along with some of their artistes, made the dual operations possible. On popular programmes such as *Ukaonu's Club*, artistes like their producer made personal sacrifices to ensure that they could feature on the programme, which was transmitted from both locations. In those days, there were no satellite feeds, and local programmes were not recorded. Live shows were the order of the day, as field cameras were not commonplace. ENG cameras were only introduced in the 1970s. The production practice in ENBC-TV was very strenuous, and the undertaking reflects the commitment of the staff. This meant the personnel involved in making programmes on both stations had to be transported from Enugu to Aba. They paid to drive across this distance of about 153 kilometres (95.2 miles); a journey of four or five hours by road in those days, travelling in their personal cars. It showed the sacrifice the people were prepared to make to keep their station alive. With such efforts, operations in Aba ran parallel to Enugu and audiences were left baffled, not sure where the staff were located. [See panel story on personal commitment]

This arrangement suggests that the investment in ENBC-TV was prudent. In spite of the distances between Enugu and Aba, it was technically possible to rely on productions from one source, while establishing relay transmitters, as WNTV did. Signals from the studios could be sent to the transmitting stations by microwave link. This would have eased the operational difficulties described above, but it would have cost the station a lot more, given the rugged, hilly terrain. In any case, the sacrifice made by the ENBC-TV staff who shuttled between the two stations allowed for closer marking of the audiences they served.

Figs 2.3 & 4: Illustrations: Station Opening Graphic – depicting the hilly terrain and the dawn, which are representations of the Enugu and the Eastern region also known as "The Land of the Rising Sun". The reference to Enugu as Coal City is a reminder of the natural reserve exclusive to the Eastern region on which its fortunes were built before Nigeria's discovery of petroleum resources. Alongside is the map of Nigeria indicating the location of Enugu state; such graphics are typically used in newscasts.

Personal Commitment to Television Service in Eastern Nigerian

Anyaogu Ukaonu had trained in America. His foray in the foreign land was enough to earn him a fan base among young people who may have secretly nursed aspirations of travelling abroad. His musical talent was a bonus. His programme, *Ukaonu's Club* run like a club for young adults, was hugely popular. It featured the latest dance moves, local talent and young people having fun. The programme had to be available on both versions of ENBC-TV. This meant that Ukaonu had to shuttle between Enugu and Aba, driving his own car, sometimes through the night after a full day's work, to enable him to meet his commitments to the audience. His was just one example of personal sacrifice.

Mariam Okagbue, a veteran broadcaster in the East, recalls how rehearsals were conducted in her home. "The job requires devotion entirely . . . everybody was competing to see who would produce the best show. It was almost like a family thing. Rehearsals [for *TV Revue*] were held in my home on Sunday afternoons." To encourage the artistes, she found that she had to hire a music teacher to tutor them. On some occasions she had to make shirts for them. In one instance she bought an artiste his first guitar. That artiste was Sonny Okosun, who went on to become a household name on the Nigerian music scene. That the artiste went on to become a national star is the sort of intangible reward for the selfless service invested in television

Interview with Mariam Okagbue July 2008

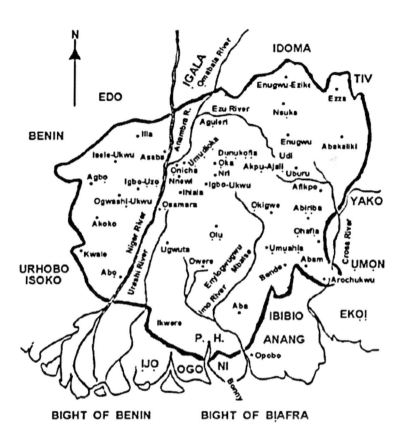

Fig. 2.5: An old Map of Igboland in Nigeria from colonial times

Audiences of Nigerian television are familiar with images of poor road networks in various parts of the nation. The problem of erosion in the Eastern part of Nigeria and the inaccessibility of certain areas is a rather familiar image. Some effort to salvage this long standing problem has been made by governments and public spirited individuals over the years. An example of such, made in recent times is the bridge constructed in 2006 by community effort. The project was led by an American engineer, Dr Todd Strong, and sponsored by the Worldwide Organisation for Women. The 80 metre-long

(264 feet) Akwugo bridge over Eze River in Ozubulu in Anambra state (images available online at http://www.flickr.com/photos/grandioseparlor/sets/72057594060534712/with/96575032/) is merely indicative of the dire state of road travel in the early 1960s and further compounded by the civil war. Inspite of the road reconstruction that occurred in the 1970s, after the war, roads leading to the hinterland, (in the east and elsewhere in Nigeria) are often just bad.

Being near university campuses was a benefit to the stations. Such association was mutually beneficial, as the academic community provided intellectual resources for the stations. The Enugu station gained a lot by tapping into the specialisations at Nsukka. The Music Department was a constant support, as students as well as staff were regularly featured. For strong debates, ENBC-TV counted on university staff from various disciplines. The publicity that the collaboration generated worked in favour of the university community, as people's expertise was made known to those who required them. This practice, which persists in the industry, is a less obvious way in which television fulfils its role of bridging gaps between different parts of society and fostering consensus. Besides the pleasurable programmes that it provides, television does a lot behind the scenes, including facilitating connections within the communities.

The management team deliberately relied on local staff to develop content for the station. They had assembled a team which had sufficient exposure and knowledge of local traditions and culture. For instance, Anyaogu Ukaonu was well travelled and popular among the young adults. His television club featured as a regular programme. Ukaonu also introduced the programme *Nigeria Dances*. With the assistance of a few younger producers, they scoured the villages to identify local dances that were available. They arranged the choreography to make the dances better suited to the medium.

Children's programmes included Mariam Okagbue's *Pick-a-Letter; Puppet Theatre*—starring (hand puppets) *Fluffy & Tiny*—and *Kiddies' Bedtime*. These programmes promoted literacy skills and the reading culture. They were aimed at broadening the scope of the children. Mary Umolu had a programme based on African folktales—*Tortoise Club*. *TV Revue* was similar to *Ukaonu's Club*, but intended for the younger age group.

The Women's Department also had its own programmes. These tended to focus on the stereotypes of women's roles. They featured everything that was assumed to be of interest to Nigerian (African) women and benefited from the inspiration of producers such as Okagbue, who had made programmes for the BBC. Some programmes hired trained demonstrators from domestic science centres to present cookery lessons. Another programme, *Beauty Box*, was all about the total appearance of the woman. This was one of Okagbue's programmes, which she sometimes presented. At other times she brought in expert hairdressers and so on. This programme was popular with men as well as women, as it set fashion standards.

The Programmes Department was administered in a way that revealed the thematic interests pursued on the station. News and Current Affairs were put together. Even from this early stage, there was evidence that the programming structure promoted a gender bias. Women and Children seemed to be a logical arrangement, as men would feel insulted to be assigned to making children's programmes. In any case, it seemed like a normal association, and roles were assigned accordingly. However, the gender politics among producers may have been less evident in the early days; because of the collective mission to establish the service, everyone was competing to ensure that his or her programme was the best. Yet this pattern of lumping women and children together was deemed by some to be demeaning to women. It was an indirect way of identifying the social status—the worth of the audience or the value of programmes. It often had implications for the allocation of resources. For these reasons, some of the women resisted this convention. Consequently, on ENBC the administrative structure in the Programmes Department was modified—women's programmes were separated from those for children. It took people of vision to insist that this be done, and even to further delineate segments within what was known as Children's Programmes. Thus there was a clear focus on teenagers as opposed to the younger age group.

Broadcasting in the early days was difficult as it meant that the producer/presenters, as the creative people were called, had to prepare programmes for both radio and television. In those days, people had to do everything themselves. For example, it took a producer to interpret properly what type of set was required for the programme.

In the event that sets had to be constructed, set constructors at that time were no more than local carpenters. Their limited exposure made it difficult for them to conceptualise ideas, so the onus was on the producer to ensure that ideas were properly translated, even if that meant they had to become physically involved in the construction of the set. This is one expression of what is often summed up as the weak infrastructure on which television was built.

Another weakness is evident in the production of children's programmes. Most of these relied on the awareness of local folk tales. Though these tales abound in Nigerian communities, few ready-made publications documented them. Often producers had to begin their research for programmes from scratch, relying on the oral tradition. The weakness of the publishing industry meant an increased burden on the broadcaster. This is another example of the infrastructure required for broadcasting.

Speaking about children's programmes in particular, Okagbue noted that, "In those days education was not such a formal thing. It was informal but in the informality there is a lot of formal training and teaching, which television houses can do." Children were used to being gathered together to listen to stories—folk tales from which they learnt moral values along with other practical lessons. This was a fun way of learning that was adaptable within the medium of television, as was proved by _Tales by Moonlight_, which originated from television service in Eastern Nigeria and still airs on the NTA network.

Following the British model, television was to inform, educate, entertain and facilitate networking with various communities. Producers scoured the hinterland for talent, festivals and dance troupes, performers whose acts could be presented on the screen; from this, they found that there was enough cultural variation in the region to keep the station busy. There was no shortage of issues to be discussed, and for children there were stories to be told, lessons to learn and new habits to promote the reading culture—but the technical and human resources were stretched. The huge task of building ENBC-TV was tackled gradually. Equipment for the permanent premises of the Enugu studios in Independence Layout was yet to be installed before the hostilities of the civil war broke out in 1966, a war which dealt a serious blow to the television industry in Eastern Nigeria.

Mariam Okagbue recalls that the Enugu station was worst hit.

Staff lost documents relating to the programmes, and though the new premises had not been fully commissioned, those consoles that had already been installed were vandalised.

> The war was a destructive experience–wanton destruction of facilities. The Enugu station was not completely deserted during the war . . . some staff stayed behind (Onitsha boys like John Chukwura) but there was not much they could do, as federal troops were all over the station. The day the first bomb fell in Enugu, I was in the Radio Nigeria recording studio. . . It shows you what danger is. The whole Broadcasting House at Onitsha Road was empty, except [for] some of us who ran further inside to hide ... we all survived.
> (Mariam Okagbue interviewed in July 2008)

The Enugu studios in Ogui Road were being moved to the designated permanent premises in Independence Layout at the time war broke out but operations were not suspended immediately. Transmission continued under cover, even as bombs were flying; staff endured until they were sent packing by the advance of federal troops. In the confusion that ensued, some were deployed to radio, which became known as Radio Biafra. Whereas the radio service of ENBC survived, television lost its prominence, at least in Enugu. Television was not as adaptable in a war situation. For security reasons, transmission could not continue. The challenge was great for both reception and production. Programmes could not be produced outside the studios, because television equipment was not portable as radio. So while Radio Biafra became a vital tool in the conduct of the war, television service was suspended till the end of the war in 1970. The Aba station managed to function for a while but television viewing was hardly suitable for people under cover in the forest. Umuahia was the last Eastern stronghold during the war. By the time Umuahia fell, the normalcy in daily life required for television reception was gone; people were too busy securing their lives to care for television. With the fall of Umuahia, television in Eastern Nigeria also fell, at least for a while.

This experience helps to reveal what is at the core of television service. It appears that while radio can be operated strictly as a channel of information, television requires a few more trappings. In spite of the other functions it performs, television is primarily a medium of entertainment. It requires a certain level of physical infrastructure, electricity supply and means of communication and transport, so that the link with the communities for its content can be maintained. It also requires stability. All these were not possible in a theatre of war, which is why of the three regional ventures, television in the east was most affected by the hostilities. The suspension of ENBC-TV, like the suspension of BBC TV (and other European stations) during the Second World War, suggests that television is a peace-time medium. At the end of the war, there was some catching up to do. With the dedication and ingenuity exhibited from inception, the people of the East proved that they were up to the task.

By the end of the war Nigeria had adopted a 12-state administrative structure to replace the old three-region structure. Much of the former Eastern region became East Central state. The Eastern Nigeria Broadcasting Corporation was thus transferred to the new East Central state government and became known as East Central State Broadcasting Service (ECBS). Though radio service resumed quickly, it took about five years for television service to resume. Education remained the cardinal task set for television, but in practical terms it was still a tool for communicating government activities to the people.

Television service resumed simultaneously in Enugu and Aba around 1975. This was a fortunate move, as it meant that when more states (that is, units of administration within the federal structure) were created the following year, the two new states carved out of the former East Central state had infrastructure in place to commence television service.

The ECBS studio in Enugu was quickly replaced and the station became known as NTV Enugu. It was well-equipped, benefiting from the largesse of the federal authorities and the administration of Ukpabi Asika in East Central state, which invested heavily in the resuscitation of the station. The groundwork for the takeover of existing stations was being laid. Asika was interested in broadcasting, even though as sole administrator in a military, rather than a civilian, regime, it may seem that he did not need to court the populace. Still, like many political office-holders, he appreciated the media and the opportunity for publicity. This was further confirmation of the perceived importance of television in the business of governance. *Face the Nation* was a programme that enabled him to give an account of his administration to the people of his state. Aba also became a fully fledged station in 1974 after the war. Asika, as the sole administrator of East Central state, was faced with the dilemma of the request of the Imo people for their own station. They were willing to pay for their station; thus, the station was resuscitated as NTV Aba and positions were filled as Imo people had made good on their promise to make contributions towards the establishment of their station.

The collaboration between state governments and federal authorities in these stations was short-lived; it was undermined with the establishment of the Nigerian Television Authority (NTA), through which the federal government brought all existing television under one authority. The loyalty of the stations was thus diverted from the state government to the federal authorities, and state governments were quick to replicate stations which gave them the autonomy they once had. In Anambra state the government conceived the idea of establishing its own station. This project was initiated by Asika but completed by Jim Nwobodo. Once it was under way, the new initiative became the priority of the state government; there was little point investing in a federally owned establishment. The Anambra Broadcasting Service—ABS (later known as Anambra Broadcasting Corporation, or ABC) emerged in Enugu, while the Imo Broadcasting Service was established in Aba in 1976.

As a medium for information, the news service was central but it is in this aspect of service that the limited capacity of ECBS-TV (later NTA Enugu) was most glaring. The station had been established without access to visuals to support its news reports. Studio productions were prevalent in television production in the early days. There was no technology to support field productions, and news reports relied on footage from international news agencies like the British company Visnews, making foreign news more visually alluring. As the station was established without processing equipment, footage of local news had to be processed abroad, causing up to a three-day lag in the reporting of events. The cost and complexity of the process meant that much of the news was foreign, and what local content there was lacked the visuals that should have been the advantage of the medium. [see panel story below]

Local talent fighting Technological Dependency

It is baffling that Nigerian expertise was confounded by foreign technology but this would not be for lack of trying. The story is told of a film-processing system that was devised locally. The occasion was a state visit, the governor of North Western state visiting his East Central state counterpart. A grand reception had been planned and the station had been challenged to ensure that its report was accompanied by pictures. An ingenious team led by Boniface Ofokaja managed to live up to this challenge with their makeshift local technology. The pictures were somewhat grainy, but the state administrator (governor) did get his pictures.

"On the first day we had six minutes of the visit, [Ukpabi] Asika was there with his guests. . . the visit was well covered. Anything done before 2pm was on the news. We created a processing system. The environment was hostile to development. Nobody really cared. The system could have been further developed and patented. It lasted a year. The technicians/ photographer were keen. If only we listen and encourage junior workers, we may have found solutions [to more problems]."

Interview with Boniface Ofokaja – July 2008

The lack of funds compounded the situation, with Enugu handicapped in this regard, as commercial activity in the city could not compare with that which made commercial broadcasting viable in Western Nigeria, in places like Lagos or Ibadan. Later, broadcasting in the Eastern region had to devise means of tapping into the thriving commercial centres in Onitsha and Aba.

On return to service after the civil war there was a deliberate effort to make television appealing to the audience. As production became easier and technology had become more flexible, it was now possible to record programmes.

When the stations were resuscitated after the war, local dramas flourished in the Enugu station. Many of the performers were not even trained in acting but they thrilled their audiences nonetheless.

One example of a successful TV drama was James Iroha and Chika Okpala's *The Masquerade*, which went on to become a national favourite. Another was *Icheoku*, with its famous interpreter, Lomaji Ugorji. This programme began as a radio show, *Nwelekebe*, and was later adapted for television, and made its mark as a network programme. In spite of the upheavals in the fortunes of the station that would be "Second to None", it can still be remembered for its contributions to the success of television broadcasting in Nigeria.

I. II.

III. IV.

V. VI.

Figs 2.6 (I-VI): Graphics showing programme titles of some popular programmes on the Enugu station. Note the social message in caption VI. Courtesy Peter Achebe, NTA Enugu

The Northern Colossus – Radio Television Kaduna

Radio Television Kaduna (RTK) is the third of the pioneering television services in Nigeria. It started as an arm of the Broadcasting Company of Northern Nigeria (BCNN), at a time when constitutionally broadcasting was under the concurrent list duties.[3] As in the other two regions, television was introduced to Northern Nigeria at the behest of the politicians, an acknowledgement of the power of the media. In this case it was Sir Ahmadu Bello—Premier of the Northern Region and Sardauna of Sokoto—who championed the cause. Being a title-holder in the most influential traditional political entity in Northern Nigeria as well as the Premier of the region made Bello a custodian of modern and traditional political authority; his endorsement gave dual legitimacy to television. In his station, the distinct nature of the symbiosis between the political elite and the broadcast medium observable in Northern Nigeria can be explained.

Based in Kaduna, the regional headquarters, the station was launched on the 15th of March 1962. Television in the North was named after the radio service—Radio Television Kaduna. As had been demonstrated in the accounts of television in Western and Eastern Nigeria, the medium, though still attractive, was the weaker relative in the broadcasting deal, compared to radio. Considering the economic state of the country and its people, there was good reason why television played second fiddle to radio in those early days. When the cost of receiver sets is set against the earnings of the average person at that time, television was beyond the reach of most. It could not compare with the transistor radio, which was very popular, especially among the rural nomadic population in the Northern region. Television's most likely audiences in the North was office workers, civil servants whose lifestyles and homes could support television viewing. Yet at the time TV sets were priced out of their reach as well. To make the service more viable at its inception, these civil servants were given loans to purchase receiver sets. Later some viewing centres were established in densely populated (low-income) urban areas.

Mohammed Ibrahim, a retired director general of the Nigerian Television Authority (NTA) recalls

These [viewing centres] were all over [not restricted to the North]...
This was a deliberate government policy that since many people, the
majority of Nigerians, could not afford the TV sets, it is better to
have either a public hole or a place where people normally gather and
they set up TV viewing centres, with an operator, a generator, so that
when the service comes on it [television and the generator] is switched
on, and when the service goes off, it is switched off.
(Mohammed Ibrahim retired director general of the NTA, interview
conducted July 2008)

Such communal use of the media was not strange, though the extent
of government support given to audiences in this region appears to
surpass what was given elsewhere in Nigeria. This may be a reflection
of the economic disparities in this region, even in the urban areas,
which are assumed to be more affluent.

The priority in the North, as in Lagos, was first to pursue radio,
which was more accessible to the audience. From the onset, Radio/
Television Kaduna were designed to operate as commercial
broadcasters, just like WNTV/WNBS. The Broadcasting Company
of Northern Nigeria (BCNN) was a joint venture between the UK-
based Granada Group and the government of the Northern Nigeria.
The first managing director, Lesley Diamond, was seconded from
Granada Manchester. The establishment enjoyed a measure of
freedom in the early days. But by 1967, following the military coup
that precipitated the civil war, which was five years into the venture,
the government had bought out its partners.

Fig 2.7: Sir Ahmadu Bello, Premier of Northern Nigeria and Sardauna of Sokoto, with an aide and one of the technical partners, at the occasion marking the establishment of Television Service in Northern Nigeria – 15th March 1962

The BCNN/Granada collaboration came with its benefits. It offered the opportunity for training local staff, with extensive supervision for production staff. It was thus possible to generate a high concentration of local programmes that were cheaper than imported fare. One would have expected the new station to make the most of Granada's stock of programmes—but programmes had to be locally relevant. Staff worked across services on radio and television. There were no rigid lines of demarcation; rather, staff were expected to multi-task, and they were happy to do so. For instance, there was no need for translators in the beginning, as staff relied on their knowledge of the culture and the language. This was efficient use of labour, although with time such arrangements could no longer suffice. Another view about the motivation for establishing television in the North can be couched in terms that resonate within contemporary human-rights agenda. That is: television was premised on respect for freedom of speech. Though Sardauna and the government needed to reach the various peoples of the North, they also needed a means of feeling the pulse of the people. Television was meant to be the voice of the people to speak to and among themselves. When the patterns of access to the medium and the directionality of communication are considered, it is safe to conclude that these were noble intentions that may have materialised only partially. The fact that service was limited, excluding much of the vast area in the region such as the northern Yorubas in Ilorin on the southern fringe and their neighbours in the Benue-Plateau area suggests this.

Though Radio Kaduna reached far beyond the Northern region, the television service was rather restricted, for technical and strategic reasons. When the television service started in 1962, it was limited to Kaduna and Kano states. Ostensibly, the reason for the slow pace of expansion was the costs involved, yet it may have been due to the need to forge unity in the region. That sentiment had been expressed earlier with regard to the introduction of wireless radio. Larkin (2008: 67) noted that Northern colonialists were worried that secessionist tendencies might fester once the strict control of wired radio was lost and audiences had a choice of stations to tune in to. For example, they feared that the Yoruba-speaking people of Ilorin area of the northern administrative region would tune to the West and be swayed by ethnic affiliations. This would have undermined the northern hegemony. They recognised even then the importance of providing relevant service to minority groups in the North. When it was introduced, radio covered the whole country and beyond, reaching Niger, Cameroon and other West African nations. This lesson rubbed off on the television service in spite of the limitations of the coverage area and programming challenges.

Fig 2.8: Map of Nigeria Showing NTA Transmitters and Primary and Secondary Coverage Areas

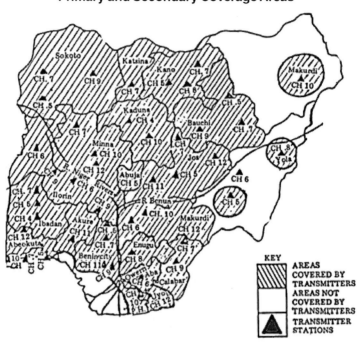

Source: NTA Handbook 1981

Effecting Coverage in a Vast Region

Transmission was poor in many parts of rural Nigeria, particularly in the North. This remained the case even in the 1980s, more than two decades after the introduction of television in Northern Nigeria, as the map indicating the location of television transmitters shows. Although by the 1980s, when RTK had been absorbed into the NTA, as had the other pioneering stations, television transmission reached the far extremes of the country, effective coverage was estimated to be reaching only 50-60% of the population (70% of the land mass), in spite of the extensions to television service in the regions. The gaps evident in the 1980s are indicative of the dire situation of coverage in the 1960s. The Northern region was by far the largest of the three regions of the federation. Its size alone posed a challenge for television transmission. RTK service was limited to the dense urban populations of Kaduna, Zaria and Kano; these were major commercial and educational centres in the North.

Kano had been relevant for centuries, being one of the stronger Hausa states, and part of the Trans -Saharan trade route. Yet Sokoto in the north-west and Maiduguri in the north-east, though significant in their own right, were too far out to be covered by the available transmitters. Sokoto was the centre of the powerful caliphate, centre of the Islamic religion and scholarship. Similarly, Maiduguri was the centre of the Kanem Borno empire in the Chad basin.

Settlements of other groups to the south of these, such as the Jukun, Igala, Tiv, Nupe and some of the other Hausa Bakwai states— Ilorin, Gwari and Yauri—were not covered by the RKT television service either (Ajayi: 2000; Falola & Heaton, 2008 give a description of the ethnic composition of early Nigerian states and societies).

The vast distances between these settlements, and the inadequate infrastructural support, meant wider television service was not feasible. For example, the distance from Kaduna to Sokoto is about 370 kilometres (270 miles); when driving, however, one needs to cover up to 425 kilometres (265 miles) to 462 kilometres (287 miles) because of the road network. The journey to Maiduguri from Kaduna is even further; a distance of 641 kilometres (398 miles) as the crow flies but between 737 kilometres (458 miles) and 801 kilometres (498 miles) by road. (Nigeria distance calculator http:// www.globefeed.com. These distances do not lend themselves to the commutes made by the pioneers of television in the east, so these cities were left out of the efforts by RTK. In any case, audience size, and the negligible prospects of adverting revenue, may not have justified the expense required to extend coverage to these places.

The area known as Kaduna state (since 1987) is just a fraction of one of the six states carved out of the Northern region during the 1976 state-creation exercise. The awkward configuration of this state made coverage from one source rather difficult. There were transmitters in Jaji to the north and Kachia to the south of the Kaduna metropolis. In 2008, for more effective television coverage, Kaduna state required four stations. In addition to the original station in Kaduna, there were stations in Zaria, Birnin-Gwari, which is further from Kaduna; and Kafanchan. An old Kachia relay transmitter had been moved to Kafanchan from Kaduna. (See map of Nigeria for example at htpp://www.onlinenigeria.com/ MapPolitical.asp or http://www.onlinenigeria.com/ MapTransport.asp).

For the station in Kano, there were three transmitters covering Kano (and the area now known as Jigawa state). When the decision was made to expand television service, Dutse (capital of Jigawa state) had no transmission at all, so it had to get a new transmitter. Arbus had a transmitter, and therefore the new station established there did not require a new transmitter. This is evidence of the deficiency of the coverage available in the early days and of efforts to correct this over the years with the creation of states. Communities in what had been another part of the north, Benue-Plateau state, for example, can now be found in four states: Benue, Plateau, Kogi and Nassarawa. These had no television service till 1975, when Benue-Plateau Television was started; even then, the service was limited in its coverage. Coverage only became more effective and new places penetrated by television signals in the new millennium with the acquisition of transmitters. The transmitter in Katsina Ala that facilitates a reach up to Wukari, then to Jalingo, helps to link different states within the region from Benue to Taraba states. Without such infrastructure the regional cooperation and the economic development that television was meant to foster could not be realised. Transmitters have since been redistributed for more effective coverage, prompted by need and by the fact that the nation could afford to invest more in these. However, with RTK, these were mere dreams.

Television was intended to be for public information, following in the tradition imbibed from the colonial masters. Government information and educational broadcasting were to be the priorities. The medium was used to speak to the people but there was hardly room for the different groups to speak with each other. The news, for example, was broadcast in Hausa and English. Weekly, there were bulletins in Kanuri, Tiv, Nupe and Fulfulde. This tokenistic effort was similar to the situation on federal radio in Lagos (up till

1978, when national radio was reorganised). It attests to the gross inadequacy in the station's attempt to reach distinct ethnic groups in the northern region. The quality of expression given to the groups within this weekly arrangement is questionable. The north was the largest of the three regions in Nigeria at the time, and, contrary to the common misconception, it is not linguistically homogenous. Uche (1998) explains that among other factors, the Islamic hold on the various peoples in the region, consequent to the Usman dan Fodio conquest of the Hausa states (Fulani jihad discussed in Chapter 1), led to this mistaken belief. However, the key to the perceived unity in the region is the wide adoption of the Hausa language, rather than Islam as a dominant religion. Hausa was widely spoken in the region, along with the other distinct languages. It remains the lingua franca in this part of Nigeria.

RTK had news programmes in other languages. There were drama programmes in Kanuri, Tiv, Nupe and to some extent Fulfulde. These were scheduled either weekly or twice a week. In addition to the English-language programme, these programmes were in the minority, as the bulk of the programming was in Hausa. It was, nonetheless, an attempt to reach these groups. In any case, the service was still largely restricted to the urban areas, even till the time that NTA was set up. The predominantly Hausa service thus appeared adequate.

Inspiration and Station Culture

RTK evolved a unique culture of service, which in Northern Nigeria has been attributed to the ideals of the man who bore the inspiration for the project. These are worth examining, given how influential broadcasters from the north went on to be. Many of them have moved around within federal broadcasting organisations such as the Federal Radio Corporation of Nigeria, Voice of Nigeria and the Nigeria Television Authority, carrying with them values and culture of operations that they had imbibed from RTK. The fact that the men from RTK rose to the helm of affairs, with several of them becoming director general in these organisations, may have paved way for the entrenchment of a northern hegemony. According to Ladan Salihu,

RTK is an institution which has refused to die sentimentally but has been long dead structurally. . . [it] is an institution that lived to champion the cause of journalistic ethics and personified same. It was taking stories from all the political parties at the time (NEPU, NCNC, UPGA) though it was set up by the Sardauna, and the preponderance of opinion was from the government of the day. That government did not prevent it from covering issues that concerned other political parties. The station gave a level playing ground for all.'

(Ladan Salihu ex-NTA News; Zonal Director Federal Radio Corporation FRCN – North West - Interview conducted July 2008)

One account of the circumstances that led to the establishment of RTK tells of a perceived snobbery experienced by the Premier of Northern Nigeria during a visit to Western Nigeria. It is said that this experience helped Sardauna Ahmadu Bello develop a high regard for balance in news coverage. Though his meeting with Awolowo was given coverage, the fact that his arrival in town was not celebrated on the television station was regarded as a slight. It would never have happened to the Sardauna in the north, where, as Premier, he held sway. Right there he asked his aides to inquire about the costs of setting up a television station. It is thought that the incident taught the northern leader that it was expedient to invest in a station that would yield to his influence and share his values.

The Sardauna's personal attributes and his conduct in office demonstrated his concern for the welfare of his people, including the poor rural dwellers. He was known to be less concerned with the perks of office. Rather, his reputation was built on his compassion, service to humanity, dynamism and good humour. These are the high ideals that he expected in the broadcasting organisation he established.

There are anecdotes about the Sardauna's acts of kindness during his many tours of the remote areas of the north. Many benefited from the contingency fund that he travelled with: women in labour, young mothers whose husbands were away on trading expeditions; villagers who had been affected by flooding. Given the difficulties in the terrain and the vast distances between settlements in the north, as described earlier, remote rural dwellers required the intervention of some external benefactor. Such a benefactor would attend to, or at least call attention to, their plight. This was rooted in the concepts of duty held as part of the tradition in most parts of Nigeria. Though television broadcasting could not disburse funds, it could bring the plight of people to the attention of those who controlled purse strings. This duty has since been internalised as an integral part of television service in the north, as exemplified by contemporary services on NTA stations in Zaria and Kano.

Following the Sardauna's leadership style meant having high regard for all people, irrespective of their creed or culture. This was a good ethos for broadcasting, and RTK sought to build its reputation on this in the 1960s. Even then, moral guardians had been concerned about the corrupting influence of the broadcast media in particular. Due to costs and technical considerations, television may have been constrained in expressing the intentions underlying the broadcast service in Northern Nigeria. We can, however, only infer from the radio service of RTK what television might have been used for. Radio Kaduna had correspondents around the country to give a northern perspective on a wide range of concerns. Its scope of coverage attracted a strong audience base. The television arm of Radio Kaduna was bound to benefit from this reputation and the relationship with audiences because of the synergy within the institution.

RTK was a regional outfit tailored to project regional interest. The station helped to mobilise support for the national forces during the civil war. The programmes were aimed at boosting the morale of the forces and their relatives, so the soldiers did not feel they were fighting a useless war. Various programmes were tailored towards explaining why it was necessary to keep Nigeria one. Programmes broadcast between 1966 and 1970 were deployed to this task but the elitist nature of the television service meant it did not have as much impact as radio. Still, TV programmes were made to reflect the local culture. This helped to homogenise the culture while reflecting the social context. The high proportion of programmes in the local language meant that outsiders to the northern culture struggled to appreciate the value of the service. Yet the high proportion of Hausa-language programmes may have helped or compelled settlers from other parts of Nigeria and abroad to assimilate within these dominant northern cultures. However, the impact of such programmes on the more likely audience is questionable. Among those who would have been young viewers of RTK in the 1960s— those who should have been the most impressionable—the memory of television reflects other priorities. As in the west and the east, the drama and musicals were popular with young people. *In the Mood,* featuring Sonny Okogwu, was one locally produced musical programme that some youth remembered. Among the local dramas which endured long enough and made their way to the national network were the comedies in pidgin English *Samanja* and *Jagua*. Like the British *Dad's Army* (1968-1977), *Samanja* is a comedy which reveals the culture of military life; the misadventures of other ranks and the frustrations of the officers. Samanja, derived from the word sergeant-major, was the central character in the programme whose misappropriation of English words was as central a comedic device as his folly. The programme reflected the dominance of the military in social life since the military incursion and the hostilities that followed in 1966.

Alternatively, *Jagua*, like many of the local comedies, focused on the domestic dilemmas of the low-income group. *Jagua*, performed in the tradition of the Yoruba travelling theatre groups (*alarinjo*), with dancing interludes, also rallied the various settlers in the north. Like *Alawada*, the programme was subversive of the patriarchal authority in a subtle way, with constant reference to the central (male) character as a foolish man, while the wife was more industrious and more perceptive.

The Beverly Hillbillies was one of the popular foreign comedies on air. Later there were shows like *Sha na na* (1977) and *The Love Boat* (1977-86). Perhaps due to the limited hours of programming or the limited number of programmes that interested them, audience recall of programmes was hazy; the impression from audiences interviewed was that the programmes were few. This may be a reflection of the pattern of viewing.

The orientation at RTK was to promote northern interest and defend northern values. That meant promoting the Hausa hegemony, protecting political institutions and the political interests of the north as a dominant force, and the power equation, the distribution of national resources. It helped to present the north to the other parts of the nation and the world. It also promoted unity between the different entities in the north. In all these, the radio arm was more prominent. In the 1960s there were common aspirations, and opportunities for different parts were meant to be equal. The Sardauna was a unifying force, as he did not discriminate, but as was shown with the distribution of transmitters, economic and technological factors meant that access was not at all equal. In any case, the politics in the north has changed since the era of the Sardauna, and the role played by television in fanning the embers of discontent is yet to be explored. As contradictions in experience and opportunities became more apparent, the unity of purpose in the north was undermined. It is only reasonable to surmise that television, along with other media, may have played a role, even if inadvertently, in unearthing the differences. It can be argued that disparities between mediated experience of life and people's physical condition is a basis for agitation. With more people becoming better educated and economically able came the agitation for the creation of states among the ethnic minorities of the Northern region. With the creation of states, the incentive to contribute to common funds to support an amorphous station like RTK was no longer there. The new states would rather establish stations that would serve their particular interests. Yet the RTK caucus remained strong even after the organisation was subsumed under the Nigerian Television Authority in 1976. Perhaps the resilience of members of the RTK system in the new national television authority attests to the quality of their training or the superiority of their mentoring system and success plans. It may also be that products of the RTK system have been assisted by the wider national politics in maintaining a measure of dominance in the NTA, as events discussed in later chapters will show.

The Lagos Service: From NTS to NBC-TV

The federal government was the last to establish its own television station which commenced transmission in April 1962. Hitherto the audience in the federal capital, Lagos, had been served by WNTV.

Lagos was in close proximity to the Western region. Unlike the earlier stations which allied with British companies, the Lagos initiative was a joint effort between the federal government and the American NBC International. Situated on Victoria Island, by the shores of the Atlantic, the station was known initially as Nigerian Television Service (NTS). This collaboration lasted five years, after which the federal authorities bought out their partners. In 1967 NTS was integrated with federal radio service and became known as Nigerian Broadcasting Corporation-Television (NBC – TV). The station had a small but strategic reach; its designated market area was then the seat of federal government, the commercial nerve centre. This city had a thriving sea port and airport, and the rail service that served the Nigerian hinterland began here.

Lagos was the base for missionary and educational activities, it was very cosmopolitan, and with its expatriate community and large population of well-travelled people, including descendants of returned slaves, a Western lifestyle was prevalent. In any case, radio, established ahead of television, had helped to introduce Western agendas to local audiences; through the radio signals and the power of their imagination, many people had been transported to novel places at home and abroad. Thanks to its strategic commercial and industrial location, the Western Nigerian television service had tapped into this market since its inception in 1959.

NTS was the late arrival on the mediascape; the newcomer had competition from the outset, but it also had a lot in its favour. In this example, benefits of a synergy between radio, television and wider socio-cultural context are apparent. Lagos had the strong legacy of a community of indigenous radio broadcasters. In fact, it had the training school for broadcasters, catering primarily for on-air presenters but also rearing producers. That it was so close to Ibadan meant it attracted some of the staff from the premiere television station, WNTV. Lagos was blessed with a thriving social scene, with a good mix of the expatriate community, the new Westernised middle-class Nigerians, dynamic Christian organisations like the Young Women's Christian Association (YWCA) and, within the club scene, musicians and a bubbling nightlife. From these, the station could draw on the performers and artistes to sustain audience interest. It also extended into many homes ideas and values that would otherwise have been the exclusive preserve of the elite. Culture was thus democratised and aspirations were fuelled.

Unlike the earlier stations, NBC-TV was in separate premises from the radio service. The Victoria Island premises in later years became the site for the operational headquarters of the NTA when it took off in 1977. Channel 10 was 'The Station for the Nation.'

In spite of the preponderance of foreign programmes such as *I Love Lucy* (1951 - 57), *Bonanza* (1959 - 73), *Perry Mason* (1957 – 66), *Mission: Impossible* (1966 – 73), *It Takes a Thief* (1968 – 70), *Mod Squad* (1968 – 73), and the dominance of news programmes, NBC -TV must be remembered for some of its local programmes. Notable among these were women's programmes such as Frances Adebajo's *Feminine Fancies*; and children's programmes—*Children's Playtime* with "Aunty" Ebun Okesola and team, including the hand puppet. There were also local variety and light-entertainment shows and of course drama programmes. Saturday nights were brought to life with Rosemary Anieze (Nigeria's vivacious beauty queen—Miss Independence) and later Art Alade (who was no less mirthful) on the *Bar Beach Show*. The show featured live bands and entertainers from the streets of Lagos. The station also had audience participation programmes: quizzes, school choirs, debates and so on, in the tradition that had been established on *radio* and the earlier television stations. These organised local secondary schools into leagues, guaranteeing audiences, and promoting healthy rivalry and a new way of life.

Sunday evenings was the slot for Segun Olusola's *Village Headmaster* produced by Sanya Dosunmu and then Tunde Oloyede. This drama series was set in a fictional but recognisable Nigerian location—Oja village. It was the highpoint of family viewing, scheduled for early evening, and ran from 1968 to the mid-1980s. The show portrayed a range of domestic and social dilemmas stemming from the need for a re-alignment of old ways to new lifestyles endorsed by Western education, hence the headmaster's prominence and the superiority of his counsel despite the presence of traditional political structures. According to the original headmaster, played by Ted Mukoro, it was a testament to the teachers and civil servants who took over from the colonialists. This was a unique production for its time, in its concept and in the commitment and dedication of the entire cast, who were required to rehearse three times a week before their performance. Until the 1980s there were no field cameras, so all productions were studio-based. The 50- or 60-minute productions were recorded in one straight take, and there was no opportunity for editing because of the recording format employed. Rehearsals and performances depended strongly on teamwork; each person had to avoid errors, because a mistake in the 48th minute meant starting all over again when the episode was recorded. There was no such luxury when it was a live show; no mean feat, as there were on average about eight cast members featured each week, reflecting a cross-section of socially recognisable types and exhibiting the nation's ethnic diversity. This was federal character in action, long before the term was officially coined.

The regulars included the headmaster, his wife, the schoolteachers, the shopkeeper/patent medicine man and his daughter or other dependants; the ruler, his chiefs and the local bar owner (pub landlady) who was also the village gossip referred to sometimes as the radio. Some represented the dominant ethnic groups: Kabiyesi (Dejumo Lewis) Chief Odunuga (Comish Ekiye), Chief Eleyinmi (Funso Adeolu), Councillor Balogun (Wole Amele) represented the Yoruba, Garuba (Joseph Layode) one of the teachers was Hausa and Okoro (Jimi Johnson) the shop keeper in the later series was Igbo (he was only introduced after the civil war). Others came from the minority groups of the Niger Delta. Bassey Okon (Jab Adu) the shopkeeper was Efik; Sisi Clara (Elsie Olusola), the headmaster's wife, was, Itsekiri; Amebo (Ibidun Allison) the bar keeper and 'local radio' put Ahoada, a town in Bayelsa state, on the mental map long before the state was created. However, it did not explore the politics of these areas through the characters, though it may have associated them with particular character traits.

Members of the cast had day jobs; acting was merely a hobby for them, yet each week the show went on, sometimes live. When there was an unavoidable reason for a central character missing a performance in the event of an emergency to themselves or their close relatives, the show could not go on. That was the case when the original headmaster (Ted Mukoro, an advertising man) turned up at a pharmacy on the other side of the city shortly before the scheduled transmission time, in full costume and make-up, in search of medication for an emergency in his family. Unbeknown to most of the audience expecting their favourite show that Sunday, *Village Headmaster* was not to feature that evening. This disappointment would have been averted had the show been pre-recorded

The experience of *Village Headmaster* also shows another challenge of programme production that earlier accounts did not capture. This is to do with the poor funding. It is reported that the programme idea was initially turned down for lack of funds. Interviews with the actors show that they were poorly remunerated; despite their passion for acting and their loyalties to their audiences, when conflicts in obligations arose, there were times when other loyalties had to take precedence. So the first headmaster, Mukoro, left the show on account of the pressures from his advertising career and personal commitments. A high turnover of other members of the cast was also noticeable in the course of the programme. There were two other headmasters (Femi Robinson and Justus Esiri) before the series was rested. Several members of the original cast either left or took long absences, thus altering the character of the programme in its later years so that it appeared to be glamorising acts which it would have been correcting in its earlier incarnation. Perhaps with a more stable cast consisting of well-paid professional actors, the station and its audiences could have expected better contractual obligations. In this we learn that the success of television was due to a strong sense of commitment on the part of the performers.

By design, each of the regional stations was meant to serve much wider areas but the Lagos station, NBC-TV, had the least coverage, being restricted to the Lagos market, though the federal government had intentions to expand its services into the Western region and beyond. In reality, WNTV was the one with the widest reach due to the pattern of settlement in this part of the country and the configuration of its transmitters. WNTV was the closest neighbour; it also had the commercial orientation required to give the Lagos station a good run for its money. In those days, there was a clear distinction between the municipalities that are regarded as Lagos today. Strictly speaking, a good proportion of greater Lagos metropolis (including Ikeja, which is now the capital of Lagos state) belonged to the Western region. WNTV had a transmitter in Abafon, near Ikorodu (which is closer to Lagos) with which it could penetrate the Lagos market. Initially the Abafon transmitter was configured as an omni-directional transmitter, beaming its signals in all directions. This way, it was also able to cater for the Ijebu area in Western state. However, when the federal government announced its intention of making television in Lagos compete with WNTV, the Ibadan station turned up the heat. A WNTV report shows that the Abafon transmitter was reconfigured, making it more directional, so that 75 per cent of its signals were beamed into Lagos, thus increasing the radiated power of its signals threefold. (NTV Ibadan, 1979: 80) By this time (about 1971) the Mapo Hill transmitter that catered for Ibadan area and environs had been replaced with a new transmitter of about tenfold radiating power. By 1976 WNTV had installed transmitters so strategically that almost 80 per cent of the population of the region could receive decent signals. This followed the propagation surveys in 1965 and the three-phase expansion programme. (NTV Ibadan, 1979) Western Nigeria had more closely clustered urban settlements. By contrast, RTK was hindered from achieving the same sort of penetration as WNTV because of the land mass occupied by the northern region, as has been discussed. In the east, the story has been told of the gruelling sacrifices made to ensure that audiences in the Aba/Port Harcourt axis were served by ENBS-TV. The land mass in the Lagos territory was much smaller than any of these, yet the Lagos market was clearly so strategic that it was given priority in WNTV's plans over many of the populated areas within the heartland of the Western region. For this reason the television audiences in Lagos environs had a choice.

The justification for establishing yet another station, one owned by the federal government, is found in the composition of the population of Lagos. Its audience is deemed to have special information needs, since Lagos was then the seat of the federal government and was the commercial nerve centre. This argument was tendered again in 1982, and the circumstances in both instances suggest that the power conveyed by control of the airwaves was perhaps the strongest motivation for establishing the station. A federal television station was seen as necessary to complement the federal radio service, as the regions had broken the federal monopoly on radio and had bettered their rapport with elite or privileged audiences with the new audio-visual medium. Television was needed for the prestige it conferred and the part it played in the politics of information dissemination.

Conclusion

Television service in Nigeria in the early days (1959 to the mid-1970s) was an achievement—even though the scope of coverage was really modest. This was due to the limitations in the capacity of available infrastructure at the time. Services required for television reception were largely limited to urban areas. The television sets were expensive, available only to those on higher income or public servants who had access to the government-supported television loan facility. Both groups were characteristically to be found in urban settlements. Nigeria's population has since grown, and the urban population is much higher now than in the 1960s, when it was more evenly spread. Whereas in 2008 Lagos was regarded as a mega-city, in the 1950s and 1960s its population was modest compared with Ibadan and Kano, which were the largest commercial cities in the west and the north respectively and already established cultural centres. Yet even back then Lagos had its prospects as a different kind of cultural centre, more cosmopolitan and outward-looking. It had a high concentration of the "right" kind of audience. For Lagos and the other regional services, there was no reason to prioritise the expansion of the television service beyond the strategic urban centres. Television acquired the reputation of being a medium for the wealthy or privileged, whereas radio was the medium for the rural dweller and the common man.

In this lay the contradiction in the mission set out for the service. The rural population who should have benefited from the laudable plans to be executed through television service education and public enlightenment to improve health and agricultural practices were the ones who could not be reached without special intervention. The solution was to establish viewing centres in rural areas and among the urban poor. These were furnished with television sets, along with generating sets to supply electricity. Staff were employed to operate these. It was a most successful policy, especially in the northern part of the country, in which such centres still exist. According to a former director general of the NTA, Mohammed Ibrahim,

> . . . in Tundun Wada [Kaduna], especially when there is a popular
> drama or local football matches, hundreds of people will be there
> sitting quietly enjoying themselves in the knowledge that it is the
> only way they could have access to this [medium]. It was very, very
> popular all over the country. (Interview conducted in July 2008)

As mentioned earlier, the WNTV project in Omi Adio was an example of these community experiments. There were also community viewing centres for audiences in the Mid-Western state and other parts of Nigeria. The need was so strong that when, in spite of the success of the early initiatives, investment in community-viewing centres was suspended by the Western state government, it is on record that in 1975 a strong case was made by the chairman of WNTV to resuscitate the project. The argument was simple: having invested up to five million naira on capital assets, with additional sums exceeding one million naira annually on recurrent expenditure, it was only reasonable to provide the service that would reduce the problem of low viewership, since the potential audience far outstripped those who owned sets (NTV Ibadan, 1979: 59). In time, enhanced per capita income reduced reliance on government-funded community viewing centres, and in the new millennium there are new public viewing spaces.

Perhaps the most significant lesson from the early days of television is the fact that the federal might had been effectively challenged. The monopoly of the federal radio had been broken. All the regions had strong and far-reaching radio services, even if television had a more modest reach. Western Nigeria blazed the trail; the other regions were not left out. Within three years, each region had its autonomous stations. Eastern Nigeria Broadcasting Service – Television was established in 1960. The station that served Northern Nigeria, known as Radio Television Kaduna, was established in 1962 as an arm of its radio broadcasting company, Broadcasting Company of Northern Nigeria. This power was not intrinsic to the medium, but owes a lot to the vision and determination of the leaders who introduced it. Long before the oil boom in Nigeria, the administrative regions were able to do the nation proud with such laudable achievements in the sphere of television broadcasting that few other nations could boast.

Each of the stations discussed above (Western, including what became the Mid-West, Eastern, Northern and the Federal Territory in Lagos) set out to serve regional interests. It was a bottom-up approach to development. Though the regional interest was paramount, it was meant to be framed within the broader national interests. This would have been ideal but for the corrupting influence of parochialism, as illustrated in the role assigned for the broadcast media in the build-up to the Nigerian civil war and the conduct of the war. This suggests that without a strong coordinating influence at the centre, the pursuit of different agendas may tear a nation apart. When the interpretation of tasks became tainted by divisive regional nuances and the sectional interests within them, the laudable national objective of stations pursuing the common good was compromised. With hindsight, it is apparent that regional visions were not as broad or panoramic as the views they expressed seemed to be. There were diverse views and identities within the regions which the dominant positions failed to encompass, so within every region, there were marginalised groups. This was a socio-cultural matter, as well as a political issue. Many communities felt excluded when their language was not adopted, and to them the success of television was not complete; there was more work to be done. Such discontent is at the root of the expansion of television service to be discussed in the next chapter.

Notes

1. Though the USA was distant from the theatre of war, American television was still affected by the Second World War. Investment in the war effort diverted attention from the fledgling TV industry, which had earlier been hit by the Great Depression. Dominick states that the regulatory body, the Federal Communications Commission, put a freeze on licensing new stations during the war, and those developing television returned to the laboratories. By 1945 there were only eight stations on air but by 1950 the number had risen to 98 (1996: 263). The performance of the television industry in postwar America was remarkable: the medium had benefited from new technology developed during the war. The industry was thus able to forge ahead of other nations in programming and manufacturing. Though it remains a dominant force in the world market, it has been overtaken by the Japanese in the manufacturing sector (Abramson, 2002:1).

2. This classification, which consists of innovators, early majority, late adopters and laggards, is based on Roger's diffusion of innovation model describing the pace of accepting change.

3. According to the 1958 Nigerian Constitution, broadcasting was on the concurrent schedule of duties, which meant that both federal and regional authorities could establish radio and television stations.

Chapter 3

Proliferation of Television Service
The Second Wave (1976 – 90)

If the 1960s were when television put down its roots, the 1970s were when the first shoots began to appear. Nigeria was firmly under successive military rulers by then and the politics of governance had changed. Unlike civilian government, the military had a unitary chain of command; central authority was very strong, and the states in the federation lacked the autonomy that the regional governments had had. With the experience of the civil war (1967-70) the central government was holding on tightly to the reins. There was a deliberate effort to propagate a television service to make it more inclusive, to have a national outlook and to coax out the best from what had earlier been planted in the regions (and later the states). As well as the regional stations and the federal government station, two additional stations had been established in the early 1970s (Mid-Western Television and Benue-Plateau TV). By the end of the decade, structures had evolved to support the tendrils of a national television service, giving birth to what today is the largest network in Africa. This chapter focuses on this phase of propagation and proliferation of that television service.

Delineating Markets

Since the industry had begun as regional initiatives, television was tied to the construction of identity. In 1977, when the Nigerian Television Authority (NTA) was established, fostering national unity was the priority of the military rulers, starting with General Yakubu Gowon, then the Mohammed/Obasanjo administrations. Delineating television markets was a technical as well as a political matter, but in the final analysis it was also about the economics – compromises had to be made because of the cost of getting the infrastructure that would offer the desired reach. However, as will be revealed in this discussion of the second phase of television's history (1976-90), TV was still crucial for asserting local identities. It remained a tool of governance, with strong government control, so state governments still desired to share some of this, as is shown with Eastern Nigerian Television under Ukpabi Asika, the sole administrator of East Central state

The 12-state structure of the federation established in 1967 facilitated the big wave of television's expansion but the trend had begun earlier. In 1963, the Mid-Western region was extracted out of the Western region; a move that splintered the cohesion in the South West, revealing some minority interests in the region. This was to the advantage of the dominant political party from the north, the Northern Peoples Congress (NPC). While it was a positive political score in certain quarters, it was symptomatic of the degeneration of the politics of the First Republic. Similarly, by 1964 the marginalised people of the north, even though they were still subsumed within the region, had begun to forge alliances which betrayed their desire to break from the dominant political force. These were people such as those of the middle belt whose interests had coalesced under the Joseph Tarka-led United Middle Belt Congress and the Kano-based Northern Elements Progressive Union led by Aminu Kano (Falola & Heaton, 2008).

Not surprisingly, these states (Mid-Western and Benue-Plateau) when they were formed were the pioneers in the expansion of the television service. These were areas that were too remote (physically) to be served by the earlier stations. Besides, there was the question of their cultural orientation and the desire for political assertion, so Mid-Western state established its own station (Mid-Western TV, or MTV) in 1973. Benue-Plateau state established a station as well. Benue-Plateau Television (BP-TV) was regarded as another remarkable feat in the history of television. It was the first colour television station in Nigeria, the second in Africa (after Zanzibar / Tanzania). BP-TV commenced transmission on the 1st of October 1975 (Salama, 1978: 12)

The quest for state government television stations also featured in other parts of the Northern region, revealing the ethnic diversity of Nigeria. Kano and Kwara, like Benue-Plateau state, were carved out of the Northern region. These states also had plans to establish their own stations but the plans took a little longer to materialise. They had all but set up their stations, as equipment had been ordered and studios built when the gear shifted in broadcast regulation. The establishment of a single national television authority which subsumed all 10 existing stations in the states overtook their plans.

No doubt, having a television station was always high on the political agenda of every state. The despatch with which the young states tackled the issue of a television service, and the ambition of each station to outdo all others, were remarkable. Uche's account of the television landscape in the mid-1970s confirms that there was a rush by the state governments to establish their own television stations. Some governments shunted the regulatory process; they by-passed the Posts & Telegraph Department, which should have allocated frequencies for transmission (Uche: 1989: 64). State governments could afford to do this because they had the clout to get their way while the different federal agencies responsible for broadcast regulation resolved their bureaucratic contentions. The federal Ministry of Communications was responsible for the frequency allocation, though the federal Ministry of Information had the oversight function on broadcasting organisations and applications for broadcast licences at the time. Processes involved in establishing a television station tended to be trying. In any case state governments were the final authority within their jurisdiction area; they determined the focus and specifications of the television service, if only by the level of investment they were prepared to make. Such willpower, characteristic of the military administration, marked this phase of growth in television. All these points confirm the symbiosis between broadcasting and politics with which the exponential growth in the television industry can ultimately be linked. Two developments illustrate this point well: the first was the 1976 creation of states exercises; the other was the establishment of the Nigerian Television Authority (NTA).

The NTA was the brainchild of a military government. It was established by the federal government in May 1977, though it took effect retroactively from April 1976. The objective as set out in Decree 24 of 1977 was

> To erect, maintain and operate television transmitting and receiving
> stations; to plan and coordinate the activities of the entire television
> network; to ensure an independent and impartial service which will
> operate in the national interest; to give adequate expression to culture,
> characteristics and affairs of different parts of Nigeria.
> (Ugboajah, 1980: 32)

By bringing all existing stations under this authority, the stations were
made to adopt the common identity. So, though the stations were
meant to be the voice of the states, the stations had also become the
voice of the federal government in the states. As a matter of policy,
the NTA established stations in every state to ensure that there was
no lack of federal presence anywhere. The authority began establishing
stations in state capitals that had none, and for those whose plans for
establishing their own service were well advanced, it acquired the
facilities that they had ordered. That was the case in Kano and Ilorin
(Kwara state). Further state creation exercises in 1987, 1991 and
1996 brought the total number of states in the federation to 36,
meaning the television industry was to expand even further. State
creation was about bringing government closer to the people at the
grassroots; with a station in each state, television was able to achieve
greater penetration of the vast Nigerian market.

Television in Nigeria thus became a uniform brand, so standards
had to be uniform. Government invested in upgrading facilities to
ensure that all stations transmitted in colour rather than black and
white. It was part of the drive towards national unity after the civil
war, and the service had to be appealing. In the third national
development plan, as much as $203.6 million was allocated to
television (Olorunnisola & Akanni 2005: 103). Having television was
prestigious, but having it in colour, especially during the Festival of
Black Arts and Culture in 1977, was even more impressive. Like the
construction of the National Theatre, the National Stadium and many
lavish infrastructural projects embarked on at that time, television
benefited from the oil boom windfall that occurred from about
1970.

With the civil war behind it, the federal government had the task
of consolidating and re-building the nation. All these concerns were
translated into the social, cultural, economic, political, technological
objectives that guided the day-to-day operations of the NTA. With

its new-found wealth, the nation was able to pay for its pursuit of a form of federalism in which the centre was stronger than the fringes. This structure reversed the earlier order that had allowed regional authorities autonomy at the expense of federal government; a military dispensation offered little room for the political rivalries that had accompanied television in the early days.

The federal government had the media within its grip with the takeover of television and radio from the states. Even the press, which had a longer and more vibrant history as private initiatives, was not spared. The federal government took over two national newspapers, *The Daily Times* and *New Nigerian*. All these entities came under the aegis of the federal Ministry of Information. Military regimes did not accommodate clashes of wills or clashes of interests. Government officials, public servants, including media practitioners, like military officers, learnt to obey the last command. The populace also learnt to comply. In any case, there was no urgent need for civilians to express dissenting views, though television still remained a prized tool of governance. However, during military coups – and Nigeria has had its fair share of these – securing radio and television stations was as vital to coup-plotters as it was to the incumbent government. Along with armouries and the residence of incumbent leaders, television stations were among the strategic targets of dissident forces. Broadcasting houses were essential in reaching the masses. Even in times of relative political stability, state administrators jostled to project a salutary image of their state whether as evidence of the state's performance, justification of their claim for a larger share of the budgetary allocation (aka *national cake*) or simply to massage egos or to keep their people sweet and happy. Television, the more visual medium, was more effective for these. Under the military, the television service was expanded and government control was made more formal.

The centralisation imposed by the new television authority was crucial in establishing control within the federal system. It began as a technical project under Murtala Mohammed (when he was federal Minister of Communication in Gowon's administration in 1974-75) and the nation had experimented with Aerostat-tethered balloon technology. The idea was to facilitate national coverage with transmitters in balloons suspended from five locations across Nigeria.

It was much cheaper than satellite transmission, and the signals would have been accessible to the viewers. If it had worked, it would have had the capacity for about six television stations and the same number of FM radio stations, receivable in any part of Nigeria simultaneously. This would have meant that the different zones of the federation had the capacity to generate and transmit programmes for national audiences. It would have been much like having six parallel networks; a feat which is yet to be accomplished on terrestrial transmission. If the system had worked, the nation could have been spared the effort and resources invested in the proliferation of stations that followed the creation of new states, though this would not adequately penetrate the grassroots. Fewer stations may help maintain a national rather than sectional focus, develop a national identity and foster unity, though the challenge of projecting cultural diversity would have remained. All this seemed logical in a military administration with its unitary chain of command. However, this technology had not been sufficiently tested, and it was little wonder that it failed, but the spirit that informed the establishment of the NTA – one which sought unity in diversity—had been established.

To create a sense of ownership and bolster confidence in the stations, there was provision for state authorities to be involved in them. State governments would have a say in the selection of the zonal board members who would guide the affairs of the stations. To some extent the success of the project relied on the cooperation of the states. NTA stations took off quicker in those states that were prepared to contribute to this new investment, some by providing the land or building the premises to accommodate the stations; some by housing core staff or even providing funds to buy equipment. Many states were happy to be involved, since it meant they could at last have television in their backyards and their communities would benefit from the effort. But the unity of purpose did not last. A different orientation emerged when military leaders returned to the barracks (albeit for a brief intermission) at the return to civil rule.

Military rule had never been the plan for Nigeria, and for every announcement of incursions into government by the military, there was a plan to restore civil rule once the mission of "righting wrongs in society" had been accomplished. There are suggestions that the military regimes themselves appeared to regard their intervention into politics as an aberration, right from the short-lived regimes of 1966-67, including Gowon's regime (1966-75) up to the Abdusalam Abubakar-led regime that ended in 1999, when the nation finally returned to civil rule (Onuoha 2002: 19 – 20; Fadakinte 2002: 40 – 41; Falola & Heaton 2008: 172 -3). Onuoha argued that

> What Nigeria has been experiencing since military rule in 1966, but in particular since the end of the civil war in 1970, is a process to establish hegemonic order, or a consolidated ruling class. This is being enforced through the constant struggle for state control (ie, the subordination of the state) among the ruling class. . . composed of military / bureaucratic, political and business class. These have at various times intensified their struggle over the subordination of the state . . . particularly since the 1980s (the so-called structural adjustment programme and the transition periods) . . . these processes of struggle properly situate the Nigerian society as a society in transition. (Onuoha, 2002: 21)

This struggle informed the orientation of the ruling class to the media. Television was firmly under their thumb. Indeed, once it was established, the NTA enjoyed monopoly status while it attempted to evolve reasonable structures to strike a fair balance between its duties to the audience and accountability to government. The first instance in which the military made good its promise to restore democratic governance was when General Olusegun Obasanjo handed over power to Shehu Shagari in 1979, signalling the birth of the Second Republic. Return to civilian rule was to test the authority and further shape Nigeria's television industry.

The Power of Language

One advantage of broadcast service over other media is the effortlessness involved in receiving messages. This, however, is predicated on one condition: audiences must be proficient in the language of the broadcast. In this, the one paradox of a centralised television service was revealed. Regardless of the location of stations, the language of broadcasting should be congruent with the local language of the audience. With a centralised television authority this could not be the case for significant parts of the broadcast day.

Limited hours devoted to local programmes in the early days led stations to rely on foreign English-language (British or American) programmes. Even with local productions, a fair proportion of programmes were in English, which was the official language, along with the three major Nigerian languages (Hausa, Yoruba and Igbo) according to the 1979 constitution. The number of languages spoken in many areas meant that television stations had to prioritise which languages to broadcast in. The inelasticity of time compounded the challenge of programming. At that time, the broadcast day was limited; transmission began in early evening, usually at about five o'clock, and stations closed down at about 11pm or midnight at weekends. For a while, *Instructional Television*, which was aimed at secondary-school students, was aired in the afternoon. This pattern reflected the lifestyle of the audiences—the workday, the school day and the expected time for leisure.

The medium was meant to promote local cultures. Its greatest benefits as a teacher and motivator of social progress would be achieved when it addressed local audiences in their own tongue. Thus, as the service expanded, the imperative to use a wider variety of languages increased. It was only a matter of time before minority groups began to nurse the desire for stations designated to their areas, to speak their own languages. For example, the Oguu (Egun) speaking people of Lagos were not content with Yoruba and English broadcasts alone. Priority would be given to the ideal that stations spoke to audiences in their own languages, discussed local issues, showcased local cultures and fostered both ethnic and national affinities. Thus attention turned to increasing the number of local programmes, many of which were still in English.

English was the language of the urban elite whom television served primarily; it was consistent with the new Western lifestyle that television signified. English was also the language common to ethnic groups across the nation. Though the use of indigenous languages had begun to flourish, reaching network audiences meant English was the predominant language of television. It was the language in which the news was broadcast, for example. Such newscasts in the early evening were accompanied by translations to the dominant local language in the designated market area. Clearly, translating messages to local languages in order to reach the vast majority of Nigerians posed a challenge; translations were not always adequate. In the first instance, news translations were not as comprehensive as the main news in English. There was more to the issue, which had political implications. "The political realities, the problem of alienating one group from the other, forces language policy to take an evolutionary rather than a revolutionary approach" (Ugboajah, 1985: 89; see also Brann, 1985; Simpson, 1985).

It was only pragmatic of television to impose certain languages on audiences. By so doing, it would also appear to be imposing certain cultures on them. However, if stations had a common language base, a situation made possible by smaller designated market areas, the potential for creating resentments within communities (on that account) would be reduced. This began to happen with the creation of smaller states in 1976. In this phase of expansion, television revealed faultlines in the administrative structures, even if unintentionally.

Television may have contributed to the revelation of such faultlines in society in another way, given the technical limitations of coverage. There were political implications of the inability to reach certain parts of the nation with television signals, when some communities had to rely on signals from neighbouring states rather than their own. The situation was bad when the states in question had different ethnic, historical or political allegiances or when their fortunes (natural resources, allocated share of federal budgets) differed. It was worse when television portrayed life as being better on the other side, when particular communities seemed to be looking on while their neighbours made progress. Entertainment programmes such as social diaries showed those who were well-connected; State Visits conferred status on the communities that were favoured enough to receive them, while communities not visited were considered less relevant, immaterial and unimportant; in these was an indication of those whom time had passed by. The fact that television, along with radio, was often the official government noticeboard (carrying government directives to civil servants, schools, health workers and the populace) presented the official dimension to the social relationships portrayed. That TV news programmes featured government projects being launched or pledged may have compounded the situation. This official dimension elevated this problem from being simply a case of social alienation to being a matter of political exclusion, and therefore a cause of intense frustration and helplessness on the part of government and the governed. Such real or imagined neglect contributes to the loss of their sense of belonging, arousing a feeling of alienation and even domination experienced by the groups on the fringe. It justifies the case for devolution in television broadcasting, though the high cost for achieving this remains crucial.

Benue state in the middle belt shares a boundary with the eastern state of Enugu (and Ebonyi). At one time the poor quality of reception signals meant that certain communities around this border had to rely on service from the neighbouring state. This is not particularly unusual, as the strength of television signals varies within the coverage area. The quality of reception may depend on the efficiency of the transmitter, since its radiating power determines how far signals will travel. The distance from the transmitter station, the terrain and the existence of obstacles in the signals' path may also determine which locations within the general coverage area receive the signals. Communities far from the transmitter in their own state (usually beyond a radius of 50 – 80 kilometres, or 31-49 miles) may not receive clear signals. If such a community is closer to signals from another transmitter, they may receive such signals much more clearly. This is recipe for disenchantment if the neighbouring state has different political priorities or if its display of news and images fails to match the reality of the eavesdropping (incidental) audience. It was evidence of the underperformance of state governments or neglect of particular communities, whether this was spoken or implied by visual elements of reports. Television may thus have informed the restlessness and clamouring for redistribution of resources and creation of new states. Once these were created, the trend has been that more television stations were established.

A different challenge was posed by the cultural incongruence of television messages that relied on what can be regarded as foreign languages. With the high proportion of foreign programmes reaching the audiences came the accusation of their corrupting influence. Moral guardians advocated that stations adopt a less liberal disposition to imported programmes to mitigate this. Some wanted stations they could control at state level (and many years later as private enterprise, as will be discussed in Chapter 5). These issues had to be resolved as the industry expanded, providing further evidence that the medium is made to adapt to its particular localities even though it is shaped by the wider forces within which it exists.

All the above reveal the power of language, whether verbal or visual.

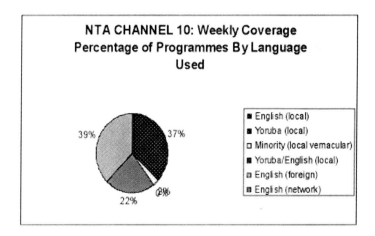

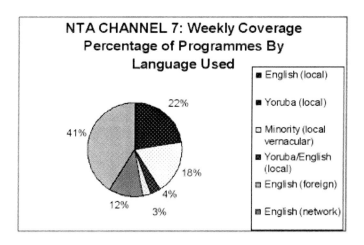

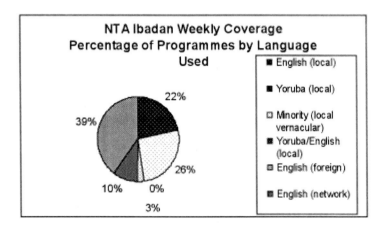

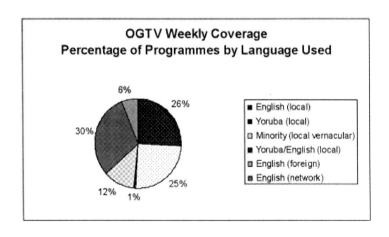

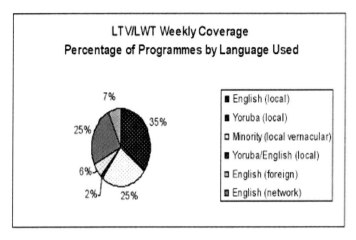

LTV/LWT Weekly Coverage
Percentage of Programmes by Language Used

- ■ English (local)
- ■ Yoruba (local)
- □ Minority (local vernacular)
- ■ Yoruba/English (local)
- ▨ English (foreign)
- ▨ English (network)

Fig 3.1: a-e Station Output by Language
Source: *Comparison of Selected Television Stations in South-West Nigeria*
1990

Standards and Staffing

One initial challenge confronting the new entity - NTA was how to maintain professional standards. From the inception of television in Nigeria, it had been possible for people to make a name for themselves with the most basic qualifications and little else to offer but a semblance of creative input, whether this was their glamorous looks or rich voices. Those in programme production, whether in News or Programmes (see Chapters 5 and 6), were able to rise through the ranks with little paper qualification if they had the creative flair – and, in the case of News, the relevant experience. Of the various specialisations in the television industry, Engineering and Accounting were the ones that required particular certification. The dilemma in recruiting adequate staff was not helped by the poor esteem in which broadcasting careers were held in certain quarters of society. In cultures that advocated modesty, TV careers were not regarded as appropriate for self-respecting women in particular. The expansion of the industry merely compounded the situation.

Setting standards was not the easiest of tasks for the NTA during the growth phase of the industry. However, social attitudes towards television practitioners were improving. More young people were charting paths in broadcasting, following their secondary education. The prospects of getting relevant university courses locally and abroad that could satisfy middle-class aspirations for higher education made such career prospects more respectable. So popular were these that enrolment in Mass Communication/Journalism courses lagged behind demand by 1978, a situation that persisted till the 1990s (Akinfeleye, 1996: 237). However, being a glamorous medium, television still attracted different shades of characters. This situation exacerbated the classic rivalry that characterises the media industry: the tension between the academically inclined practitioners and those who rose through the ranks and who, for their vocational experience, are known as professionals. The bone of contention here is the relevance of the knowledge, skills and the critical acuity that academic programmes emphasise. The debate is not restricted to Nigeria. It has been said even of Britain, that ". . . there was a perceived gap between the skills and abilities of communication and media studies graduates and the skills of people who were going to be of immediate use to media industries" (French & Richards, 1994: 91). While poor alignment of curricula with industry needs was central in the British experience, poor (or non-existent) equipment for teaching practical elements of the curriculum, and a high teacher: student ratio made Nigerian graduates appear suspect.[1] This gap within the workforce made the industry somewhat vulnerable.

Recruitment and promotion were subject to external interference. Sometimes this was merely a means of solving private dilemmas, as was done with other public corporations; at other times motives were more sinister. There was interference from those who needed to secure a sphere of influence; some had vested interests in planting moles in the organisation. Politicians and influential members of society used their positions to impose relatives and friends on the organisation. There was something to be gained by having loyalties within the station. This was evidence of abuse that the institution was subjected to, stemming from the insecurity of the political class and the mutual distrust that festers in a large organisation. Block et al (2001), citing Greiner's model of company development, point to the series of crises that a growing organisation can expect. There are five phases of growth, with particular types of crisis associated with each one. First, growth through creativity is dogged by the crisis of leadership; then growth through direction attracts a crisis of autonomy. These tend to occur with young and small organisations. The other stages of growth include those that occur as organisations respond to issues stemming from delegation, coordination and collaboration. The organisation evolves through these stages, having overcome internal crises of control and red tape. This fifth stage marks the maturity of an organisation but the nature of the crisis which accompanies this stage cannot be predetermined (Block et al 2001: 166 – 168). Strictly speaking, considering the age of some of the stations within it, the NTA was not young at the time when it was formed and even during the Second Republic (1979-83). It had experienced managers but because of the authoritarian approach of the government, it was treated as an entity that required direction. Naturally, a crisis of autonomy was to be expected.

With regard to staffing, external pressures on the organisation to be nepotistic in recruiting failed to take cognisance of the fact that television was to be a medium that would educate. To do this and make an impact, staff involved in programming needed to have a reasonable level of education. There were those who felt a university diploma was the reasonable minimum qualification to have, considering the nation's improved educational attainment. This had been the decision at Western Nigerian Television (WNTV), and the position was adopted in the new authority. Some veterans were concerned that the industry was growing up to betray the personal sacrifices which had been made in the early days. With time, the attitude towards pre-requisite qualifications has changed, especially as relevant courses became locally available through universities and the NTA TV College which was established in 1980.[2] However, the recklessness in recruitment, which got out of hand during the Second Republic (1979 – 83), must be regarded as a negative outcome of this phase of television's expansion. It took the return of the strong-willed military regime of Buhari/Idiagbon to tackle this decisively.

Their actions were also informed by the need for economic reconstruction, strict regulation and control required by the Structural Adjustment Programme. Part of this was to trim the size of government organisations.(Fakiyesi, 2005: 219 – 231).

Evolving Institutional Cultures

Establishing the television authority meant more than siting stations for effective national coverage. The hard graft of coordinating practices in the different stations was necessary if the goal of using the medium to promote unity was to be accomplished. A national institution needed to evolve. This was the task of the first director general of the Nigerian Television Authority. Vincent Maduka was general manager at WNTV at the time of his appointment as director general of the NTA. He brought to the job the proud legacy of the pioneering station, especially its high standards and its rigorously tested structure. This engineer had a vision to establish similar standards in the newly established authority. There were a few initial considerations. How do you maximise the potential of staff in an industry that relies on different specialisations? How do you avoid the excessive wage bill that accompanies a large staff? Specialisation as it existed in those days was not the most efficient way of utilising staff. This was evident in the NTA newsroom operation, which relied on news-writers from News department and newsreaders / presenters from Programmes. This system meant that the line managers the newsreaders reported to were outside the News department. In any case, the two departments had somewhat different orientations, though both had responsibility for the station's reputation. One was more laid-back and the other more disciplined by time constraints. Though the aesthetic consideration that underlined the synergy between the two departments was logical, it was a recipe for operational conflict, especially if presenters were star personalities with egos to match.

Some of the news-writers in the newsroom had come from the print media. That they could write, therefore, was not in contention; but they still had to reorient themselves to writing for television. Their scripts inevitably fell short of contemporary standards, since there was inadequate consideration for visuals. In any case, even if proper television scripts were written, there were not enough cameras to supply visuals. Television news was thus little more than radio news read by a TV personality. The least that could be expected was perfection in the delivery of the stories. For this to happen, news-writers needed the cooperation of the presenters, expecting them to rehearse the scripts. This was a sticky issue if it turned out that the newsreader had other priorities. Some newsreaders were more concerned with their looks than with their smooth delivery of the news. This galled the news-writers and created operational difficulties.

To resolve the issue, Maduka introduced a system in which presenters were given a choice to properly join the news team or remain in Programmes. This was to ensure that there was adequate training and cooperation within the news team and to enhance the quality of the authority's main output—news. In this arrangement everyone in the news team was expected to be trained in all aspects of news. Presenters were expected to do more than just read a script; they were also to be proficient in gathering news and conducting interviews. Writers, on the other hand, were to learn the disciplines of reporting, along with basic presentation. Even if the team was small, by its efficiency it could build up a reputation for the authority. The news team could form the bedrock for station audience loyalties, following the tradition of American news anchors like Walter Cronkite and Dan Rather, whose charisma, combined with their personal strength and experience meant they were more authoritative (Boyd, 2001: 164 – 5).

This move seemed like operational suicide, as most of the personalities whom the audience knew and loved had to withdraw from presenting the news. People like Mike Enahoro, Julie Coker, Bode Alalade and Taiwo Obileye disappeared from the news slot. A new generation of newscasters and television reporters emerged on account of this. Bimbo (Roberts) Oloyede and Ronke Ayuba, were joined by Sienne Allwell-Brown, Ruth Benamaisia-Opia. Elizabeth Nze, Sola Omole, Bayo Adewusi, Kehinde Young-Harry and Tokunbo Ajai. They brought a fresh, feisty orientation to news reporting. Because specialist news desks were introduced, newspeople came to be associated with particular beats. Lola Alakija and Kayode Williams were the legal correspondents. Frank Olise was known for his human-interest stories, though it is for his Sunday-night news magazine programme, *Newsline*, that he is better recognised. Others, like John Momoh, Donald Ovareghjo, Cyril Stober Sola Adelehin, became regular guests in viewers' homes. These faces dominated the screen in the early 1980s but news producers such as Richard Azoro and Gold Oruh were less well known in person, even though their names were regularly on the credits.

Thanks to the introduction of ENG (electronic news gathering) cameras, newscasts received a new lease of life; there were more "stand-uppers" in news reports; man-in- the-street interviews (vox pops) became a regular feature in the 1980s. Biodun Shotumbi, who was later to become deputy director, News, was one of the first to give live news report. He is an example of someone whose personal development on the job was remarkable, as he obtained a PhD while in service. This was an era when university-educated journalists made their mark in television news. Madu Mailafiya, who had once been at Ahmadu Bello University and later became NTA director of News, had a doctoral degree from Glasgow University. Lola Alakija and Chris Anyanwu both had Masters degrees from American universities. A range of news traditions (Nigerian, British and American) began to lock horns in the newsroom. New standards of professionalism were emerging in the Lagos-based news directorate. Yet there were many more NTA news personalities who remained obscure to national audiences because of their distance from the hub of activities. Those who were more ambitious had to find ways of getting transferred to headquarters. Being on national news meant greater recognition and enhanced career mobility; indeed, it proved to be the launch pad to more lucrative ventures for some.

Following the tradition of annual scouting competitions which WNTV Ibadan used to have, the NTA had several festivals (such as the Nigerian Television Festival, Nifest) to showcase talent and generate ideas for the different programme forms. Attempts were made to alternate the focus of attention between drama and documentary to give different departments (Programmes / News and Current Affairs) a chance to shine. Entries at such national events brought staff at the local stations into the limelight, giving them access to the better opportunities to present their talent. The prize for winning the maiden competition was a six-month stint at the headquarters in Lagos, with access to the best facilities available (after which the winner returned to his or her base). The producer of the popular drama series *Cock*

Crow at Dawn was a product of one of such competition. Peter Igho was based in Sokoto when he won the Best Drama Award at the first NTA National Competition with an entry called *Moment of Truth* in 1978. In 1982, the same production won first place at the Union of Radio and Television Organisation of Africa (URTNA) Competition in Algiers. He proceeded to have an illustrious career till he retired after many years as an executive director at the NTA, having been involved in many well-loved network productions.

Cock Crow at Dawn (1980) was a serial drama set in a northern village, focused on the return of a city-dwelling middle-class family to rural life and their efforts to succeed off the land using modern farming techniques. It promoted the ideals of the back-to-the-land project (Operation Feed the Nation – OFN / Green Revolution) which was a cardinal government activity (1976-83) (Ume-Nwagbo, 1986). The series enjoyed the loyalty of national audiences as well as corporate sponsorship (United Bank for Africa – UBA). Without the initial exposure of the competition, such talent might have remained obscure.

Fig 3.2: Cast of Cock Crow at Dawn *on location*

For his success at the competition, Igho—who, like a number of NTA staff—worked in the celluloid medium, got the facilities to produce the first two series of *Cock Crow at Dawn* in the film medium. The process and cost of developing these was burdensome but the authority kept to its side of the bargain while steering the producer in the direction of video. The argument for video was technical and financial. Video was a more stable technology, it was more flexible for television operations and it was less expensive. It won out in the end, so subsequent series of *Cock Crow at Dawn* were produced on video and the programme was no less popular. This decision shows that television managers must consider creativity from a broad perspective. Clearly, the success of the programme transcended production technology, although that does contribute to the quality of the production.

The new era of professionalism was rocked by the politics of the time. These modest gains were not allowed to take root before the acrimonious proliferation of stations began. The growth in the number of new state government-owned stations relying on the limited pool of educated or trained staff would no doubt have been a drain on expertise. It would have been noticeable in the quality of the output of the stations but the politicians were either ignorant of this or they simply did not care.

In the keenly contested 1979 presidential elections the National Party of Nigeria (NPN) emerged as the ruling party at the federal level. The winner-takes-all attitude which characterised these elections made this period a critical milestone in the history of broadcasting. It signalled the birth of state-owned stations. The landslide victory left the NPN in control of two-thirds majorities in 19 states along with the presidency. Once civilian rule was restored, states that were not under the ruling party made no bones about their intentions to start their own stations. They were correct in their anticipation that the NPN would regard the NTA as its exclusive privilege. The dispute and rancour in the run-up to the election and afterwards convinced them of the need for channels that they could control. This was a concern for the new NTA (it had only been established in 1977). Experience of television broadcasting in the First Republic had shown that politically motivated rivalry was only going to compound the battles for the professionally minded practitioners within the federal establishment; those who aimed to maintain an impartial and balanced service according to the law—the NTA enabling Decree. If the states proceeded with their plans, competition was bound to emerge. Partisan broadcasting would become the norm when stations were obliged to demonstrate their loyalty to their *owners*. Staff aspirations and battles to promote these professional (ethical / moral) standards in the NTA would be frustrated. Such professional concerns failed to gain support in a politically charged climate, in which people would do (anything to get ahead) or die (trying to do so). The view was misunderstood by the opposition party, which regarded NTA as obstructive and afraid of competition. Calls for fairness and balance were regarded by the ruling party as collaboration with the opposition. Such was the tightrope that broadcasters had to walk at the dawn of the Second Republic.

Still the NTA attempted to raise standards in political broadcasts. It produced guidelines for the coverage of elections and political broadcasts in general. By codifying what were acceptable routines and setting out clear parameters of practice, it attempted to institutionalise good practice. The guidelines were clear about how political programmes should be produced and presented, and how to engage with political parties to maintain the independence and integrity of the authority. The authority clearly outlined aspects of professional conduct that individuals, as reporters or members of production team, would be personally liable for. The authority had its own share of responsibility. For example, it was to ensure that coverage was fair and balanced; in effect that there was a right of reply. Programmes on NTA stations had to reflect appropriate cultural sensitivity. The stations were to be audience-oriented in accordance with the public-service traditions and well aligned to the varied (social, cultural, political and technological) goals of the nation. These guidelines should have complemented the official conventions of television practice, assuming that these were always in accordance with desired standards. Individuals within the institution were subjected to a lot of (external) pressures, especially in the wake of the civilian dispensation.

Maduka's administration as director general put him in conflict with certain influential politicians. His relationship with the presidential Special Adviser on Information, who was acting in the capacity of a minister as chair of the interim committee in charge of the NTA, is instructive. Since there was no board of governors, this committee was in charge and the special adviser was like a sole administrator. Maduka recalls receiving several unpalatable directives from him. This was during the Shagari administration in the Second Republic.

Chief Olu Adebanjo had relevant experience that qualified him for the role of Special Adviser on Information. He was a journalist.

In the First Republic he had worked for a partisan newspaper, The *Daily Express*, an Action Group paper. By the Second Republic he had crossed the political carpet, abandoning his allegiance to Chief Obafemi Awolowo's Unity Party of Nigeria (UPN), which was still in the opposition party, and was working instead for the NPN, the ruling party. Perhaps in a bid to prove his loyalty to his new party, he proved to be very high-handed in his role overseeing the channels of public communication. His political motivations rather than professional considerations seemed paramount. The NTA fell prey to this, and any attempt to resist the political machinery was crushed. Anyone who tried to stand against him was branded disloyal—a UPN sympathiser. Vincent Maduka was so branded in spite of the fact that even the UPN perceived his organisation as the mouthpiece of the NPN. He was caught in the party crossfire just because he tried to uphold standards that professional broadcasters believe in. He also had many unsavoury tasks to execute, directives that had to be implemented. There was the case of a reporter who had to be redeployed simply because her probing questions during a critical interview had ruffled the Minister of Justice. It was deemed an embarrassment to the ruling party, made worse because it had come from an "ordinary reporter". The reporter had to be redeployed even though it was against the professional judgement of management.

In Enugu, a general manager was removed for daring to be professional during a political "war" between Alex Ekwueme, the Vice President of the nation (an indigene of the state) and state Governor Jim Nwobodo. The general manager had been accused of blocking the use of a camera deployed to cover the Vice President's wife. Denial of such coverage meant the ruling party had less photo opportunity in a state that had dared to vote for an opposing party!

Forces within the NPN cared less that they disrupted the laid-down order of seniority considered in the appointment of general managers; their man had to be put in charge above those more senior to him. In this instance the proper succession order would have favoured a woman. This prospect was laughable to the powers that be, so they mounted pressure to put their preferred candidate in the post by any means. Such individuals were needed to guarantee that they got the right kind of publicity to secure the fortunes of the party.

To some extent, there was credence in the charge that the NTA's operations during the Second Republic were at the whims of the ruling party. Party liaison officers in the states as well as party loyalists within the organisation were used to monitor NTA stations. Individuals and incidents that were regarded as threats were promptly reported and summarily dealt with. For instance, the authority was forbidden to cover a press conference called by the late statesman Chief Awolowo regardless of the merit of what he had to say. Only so much could be done for professionalism under such politicians.

Those were heady days when politicians did not want to hear the voice of the opposition; they preferred the sound of their own voices. Three decades later, the political class, like the broadcaster, is evolving. There is an understanding that governments want stations so they can they can sell their programmes to the populace. Since television is a publicly owned medium, some broadcasters have chosen to take the view that government programmes are people's programmes. With the view in mind that, if television is a publicly owned medium, perhaps politicians can be forgiven for expecting free access to it. What, however, marred NTA's reputation was the politicians' stranglehold on the channel. Added to this is the morbid fear which rejects and prevents any dissenting views, while successive governments and pet projects of "first ladies" enjoyed unbridled and unlimited media attention. With the restrictions to access and fear of reprisals from sections of the community, including those that can be regarded as political elites, television in 1979 was nothing like the ideal public sphere for democracy that Habermas (1989) envisaged.

With the passage of time, audiences have become savvier and there has been a gradual shift away from whimsical acts of control, though the problem persists in some states. There is now evidence that clashes of interest between government and professionals can be resolved through dialogue. Strangely, the military rulers were thought to be more reasonable than Second Republic politicians. Mallam Mohammed Ibrahim served as general manager in NTA Kaduna and later as director general of the NTA. The following is his account of how he dealt with such situations.

> You discuss with the politicians. Let them know that 'we are no use to you, if we cannot be believed.' If the public does not believe what

we are telling them, we are no use to any government, so government has to allow dissenting voice[s] to come in. They have to allow the opposition to have a say so that . . . we can balance the story, so there is government side, there is opposition's side. . . There are always **two sides of the coin** . . . so generally they tolerated us – of course grudgingly – but there is nothing that anyone can do. [my emphasis] (Interview conducted with Mallam Mohammed Ibrahim July 2008)

Analogies such as *the two-sided coin* rely on familiar imagery and help to situate Western news values within traditional value systems. Although it is said that leaders are unquestionable in their authority, that their word is law, the concept of balancing viewpoints finds resonance in traditional views of justice and fair play. There is a Yoruba saying, *"Agbo ejo apa kan da agba osika"* (one who reaches a verdict based only on a one-sided account has been grossly wicked). Other ethnic groups harbour similar sentiments which show that there is nothing alien about professional news values. Even when audiences draw on the traditional paradigms, they reject skewed messages. So, as media managers warned, TV stations fail to have the impact that political leaders hope for if they are one-sided. Television managers definitely knew that audiences responded with scepticism to such propaganda. This is further attestation to the power of the audience and evidence that media power does not reside in any one location, it is contested.

There was the awareness among the more senior broadcasters in particular that their service should be to the public, not to any individual or government. Whatever was in the public interest and was considered right was broadcast, regardless of whose ox was gored; regardless of whether it was considered right for government or not. According to Mohammed Ibrahim, ". . . this is the main principle of public-service broadcasting. The public should be served [and] not any individual or sectional interest" (ibid). This is a position that permeates the rank and file of television broadcasting – although, as will be revealed, the interpretation of this may differ between individuals and organisations.

The laws setting up the various organisations have helped to entrench this orientation since they were fashioned after the British Broadcasting Corporation's model. However, in interpreting the ethos that worked for the British, organisations and individuals had allowed their socialisation experiences and peculiar circumstances to inform their view of who government is and what is right for the people. The fact that, unlike the BBC, the NTA and most other government-owned stations in Nigeria lack financial independence means they are susceptible to the tangible threats wielded by those in government. In order to survive, television management learnt the art of negotiation.

> Generally we looked at the law: are we on the side of the law? If yes, even if they [government officials] complain, if we had good reason, we had the advantage [opportunity] of informing them, telling them that 'This is not right; this is the approach you should take. You should be able to tolerate criticism. . .' I think that is why they found out that there is some element of balance on the NTA news today more than there was in our time.
> (Interview conducted with Mallam Mohammed Ibrahim July 2008)

This tolerance must be seen in the light of the shift in the political orientation from military autocracy to a democratic dispensation, and the struggle to reinstate the latter. It may also reflect the level of enlightenment among the political elite and the confidence reposed in their appointees to the post of director general of the NTA.

Though as political appointees director generals may be responsible for the nature of government relationship with the stations, they are not necessarily the government's puppets. Rapport, if not trust, which allowed professionalism to inch its way into the practice of broadcasting, was facilitated by the degree of confidence that political leaders had in their appointees. The system thus had its own advantages. As television broadcasting emerged, leaders were learning to trust broadcasters, to see them as critical partners in the process of governance and nation-building. Clearly there is a place for dialogue and a place for radical activism if broadcasting is to be responsible.

> I started as GM in a state-owned station and the problems I faced there were the same problems I confronted as DG [of the] NTA: working with the military, rigid approach, rigid rules, too sensitive to any type of criticism. But at that time a lot of them were colleagues and friends. You can sit down over a cup of tea and say: 'Look, don't think that we are trying to sabotage you, but if we don't do a,

b, c, d, e, f, g, we are of no use to you, so you have to allow us to make mistakes and correct those mistakes. That's the only way … we can develop sound relationships between you and us, between us and the viewing public generally and everybody. People speak highly of the BBC, VOA, CBS, NBC. Yes this is because of their freedom to operate within reason. That is why people admire them, appreciate them. If, however … you tie our hands, in our attempt to achieve what they had already achieved, then you will not continue to enjoy us, especially when it comes to matters of News and Current Affairs. These are the most sensitive aspects of our operations'.

I had a situation where a very senior official was sacked by the government and we were notified over the phone and expected to broadcast the information on that basis. I was then DG, NTA. My director of News & Current Affairs at the time [refused; he] said, 'No way! If you want, you can come to the studio. We'll put you on camera; you say it, so that we can have the actuality. I do not know whether it is you, the press secretary or another voice that is being manipulated but [the matter] is too sensitive [I can't act merely on the strength of a phone call]'.

Would you believe that earned him the sack? They sacked him, and sacked his assistant. Somehow we protested; [warning that] if they took the government to court, the government would lose because the staff [were right,] they acted within the regulation. 'So you [the government] had better rescind your order'. They did'.

There are many examples of cases where those in authority would ask to have news stories embargoed. At a time they would even insist on dictating from what perspective stories should be told. This does not happen now; the system is getting more mature. The leadership is also becoming more conscious of public sensitivity. Their outlook is different, their culture is different, even their capacity is different, so the level of tolerance is much much higher. I hope it will continue.

(Mallam Mohammed Ibrahim Interview conducted in July 2008) These observations were corroborated by those still in service in 2008.

State Government-Initiated Television Services

This chapter has so far focused on experiences in the federal establishment, the NTA and how its policy of establishing a television station in each state contributed to the exponential growth of the medium in Nigeria. This occurred under the military rulers, but experiences of state government ownership of television stations in the ensuing rivalry between state governments and federal authority in the Second Republic (1979-89) is explored in this section.

It was made possible only after the 1979 review of the Nigerian constitution, conducted by selected representatives of the people from around the nation. The new constitution renewed the right of state governments to own television (and radio) stations. It even allowed for private ownership of broadcasting stations but only if such private bodies were granted permission by the head of state. This was an opportunity to undermine the monopoly of the NTA; it certainly made room for adversarial broadcasting on an unprecedented scale, and the federal might was flagrantly tested. This section will document the celebrated case that opened the floodgates for the establishment of television stations by state governments.

Once civilians were back in power, they wanted stations that they would have absolute control of. Lagos was the test case for this assertion of state autonomy but to appreciate fully the events of this time, it is important to understand the process for allocating frequencies for broadcasting. Kehinde Eleshin, in his discussion of the Ogun state experience, gives a clear insight into the issues involved.

> It is standard international practice . . . that frequency allocation is always under the domain of a single authority, usually the central government of that country. The reason is obvious. If everybody is allowed to broadcast on the frequency of his / her choice, there will be chaos in the use of [the] frequency spectrum . . . By allocating different bands to these users in every country, order is maintained. The organisation that is responsible for ensuring this orderly use of the frequency spectrum worldwide is the International Telecommunication Union [ITU] based in Geneva. Nigeria is a member of the ITU and accepts responsibility for frequency allocation to the various users of the frequency spectrum within its borders. (Eleshin, 2006:122)

Telecommunication frequencies are used legitimately by a wide range of users – telephone companies, the military, maritime organisations, civil aviation as well as radio and television. Chaos in the use of this resource could have catastrophic effects, so there is legitimate technical justification for this control over the airwaves. However, this needs to be exercised responsibly. There was a clear procedure that went beyond the national authorities, as the above quote shows. In 1981, anyone interested in telecommunication would make an application for frequencies to the federal Ministry of Communication, having met the set requirements. The ministry then allocated the frequency and registered it with the ITU. Only then could a frequency be used. The NTA had gone through this process for additional frequencies as part of the preparation for the Festival of Arts and Culture (Festac '77). Three additional transmitters had been procured for NTV Lagos (former NBC-TV). International approval had been received for channels 7 and 5 in the Lagos area. These were to complement the existing Channel 10, which was upgraded to colour transmission. After Festac, the other channels were available even though transmitters had been procured. These remained in storage, since they had not been commissioned for use. In view of wider needs within the authority, the transmitter registered for Channel 5 was sent to Makurdi and the transmitter for Channel 7 was reserved for the proposed Lagos state station. That station was meant to serve the people of Lagos, as did other NTA stations in the states.

Initially the Ministry of Information, which supervises the NTA, resisted the plan for a new station in Lagos state, on the grounds that Channel 10, the Lagos station, was already sufficiently popular. Yet, as the authority argued, Channel 10 was not configured to address the grassroots of Lagos state. Unless Channel 10 was going to shift its focus from being a metropolitan station, a new station that would broadcast to these people, which meant broadcasting in Yoruba and Oguu (Egun), was justified. With this decision Lagos already had two channels and no need for a third, hence the justification for despatching the Channel 5 transmitter to a state that had greater need for it. This was the peculiar situation that preceded the federal government resistance to the Lagos state government-owned station.

Alhaji Lateef Jakande was executive governor of Lagos state in the Second Republic. He was a journalist, and on assumption of office as governor, he immediately began the groundwork for establishing a television station in Lagos. His team included staff members of the NTA, indigenes of Lagos state who had left the NTA to serve their state government; some were senior enough to be aware of the strategic plans discussed above. Lagos state was thus privy to the fact that there was a frequency that had been approved by the ITU but which was not in use. Access to such inside information may have hastened the processing of Lagos state's application for the television licence. It secured Channel 5 for its operations. This caught the federal authorities off guard but certain elements in government were determined to counter this move. The state government station was resisted on technical grounds.

Eleshin's account gives a more benevolent view of the federal government's position in the controversy. Up till this time in Nigeria, television transmission had been on VHF channels. If Lagos state was allowed to keep the secured approval for a VHF channel, other state governments which wanted television stations would expect same. The federal authorities panicked, fearing that the deluge of requests would lead to congestion of the airwaves (there were 19 states in Nigeria at the time, today there are 36). As a result, the frequency allocation policy was changed. Only NTA stations could be on VHF; states stations were to be assigned UHF channels (Eleshin, 2006: 123 – 4). Consequently, Lagos State's initiative – Lagos Television (LTV) was to revert from VHF to UHF. This seemed punitive, especially as the station had not been particularly welcome; the switch proved to be extremely costly because it had already bought the transmitters and equipment for VHF operations.

All indications suggest there was more to this than the technical concerns. It was not enough that there were two federal stations within the same market; individuals in the federal government were going to contend with the state government for this third one. This was war on the airwaves. Queries were issued to those who were suspected of collaborating with the state government. These included those who had facilitated the application for the initial licence at the ministries of Communication and Information, with little regard to the fact that they had been acting within the provision of the law. There were acts of intimidation to scare off anyone who dared to oppose the view that the new station could not have Channel 5 because it had been earmarked for the NTA.

The Lagos state government dug its heels in. This was a test case, as the new policy was detested by all state governments. It was tantamount to a denial of their right to television broadcasting.

To the layman, the main difference between a VHF frequency and UHF frequency is that the former covers a wider range than the latter. Therefore one would require more [UHF] transmitters ... to cover the same area covered by [one] VHF transmitter. For example, a UHF transmitter can only cover the township of each of our state capitals. That is to say, in Lagos State, for instance, whereas our present two VHF transmitters cover the whole state, we shall require eight UHF transmitters to cover the same area. The cost is staggering. Whereas VHF transmission would cost us 2,196,876.61 naira, UHF would cost us 10,750,000 naira. Bigger states are much worse off. For example, the [old] Kaduna state government reckons that it would require 80 UHF transmitters if it is to cover the whole state. The cost would be in the region of 100 million naira. It goes without saying that the Kaduna state government cannot afford that kind of expenditure on television alone.

(Alhaji Lateef Jakande on The Travails of LTV in LTV Next April 2007 pg 6)

Jakande's case was clear. State governments' budgets could not compare with that of the Federal Government, yet the states were expected to take the more expensive option. By reserving VHF transmitters for NTA stations alone, the federal government was reserving for itself the right to effective coverage of states. State television stations would be handicapped in covering their own states. This seemed like bullying, as other incidents confirmed.

The rather unconventional solution to retrieving Channel 5 from Lagos Television was one informed by a technicality in the law: you cannot dispute ownership without taking possession. The federal authorities pulled out all stops to secure the channel. No cost was spared to ensure that a new NTA channel was established on Channel 5. A new transmitter was flown in, and as soon as this was installed, its signals were beamed out to block Lagos Television. In the end Lagos Television was moved to Channel 8. Lagos ended up with four stations.

Professional broadcasters recognised the entire episode as a breach of broadcast courtesies, especially as it happened in peacetime. The politicians in both parties saw it as an all-out war; it was a power show; a worse form of competition than Vincent Maduka had sought to avert when he tried to talk the states out of their initiative. But this situation defied professional consideration. The states were toeing the party line.

The act of aggression further damaged the reputation of the NTA and split audiences along party lines. Audiences and those in opposition now had good reason to see the NTA as the mouthpiece of the federal government. History had repeated itself as opposition parties concluded that their best interests would not be properly served by the NTA – even those NTA stations operating in their states, in the facilities that they had paid for, as was the case in Anambra state. Because of the centralised nature of the institution, NTA stations' loyalty was to the federal government, and this meant they could not be trusted. The need to establish new media outlets removed from federal government control was clear and imminent in the states which had rejected the NPN at the gubernatorial elections. Thus began another round of state-initiated television services.

The states in the former Western region (five states carved out of the old Western and Midwestern states, otherwise known as the *Awo* or LOOBO states – Lagos Oyo Ogun Bendel Ondo) led the way again. This was hardly surprising as the political affinities and allegiances of old remained, and the drama was re-enacted. These were the states controlled by the UPN, under the leadership of Chief Obafemi Awolowo, the Premier of Western region credited with introducing the first television station in Africa. Lagos Television was merely the first to come on stream. Ogun Television, Oyo, Bendel, and later Ondo states followed soon after. Apparently establishing television stations in the states was a party decision. The UPN states acted in concert but they were not alone in the end. Up and down the nation, states that were not under the ruling party (for example, in the East, Imo and Enugu states controlled by the Nigerian People's Party -NPP) did the same. In the north, the People's Redemption Party-led state government in Kano and Plateau established television stations as well. These were the 'Progressive States'. One observer noted that

> The stations were established amidst a lot of rancour and bitterness. The politics which characterised the First Republic and fuelled the establishment and running of the early stations was still in existence. The awareness of the television stations as an important tool of

> governance had become so prevalent that no government attempted
> to function without it ... Just as the press gave African nationalism
> its primary means of dissemination, so today along with radio and
> television it gives politicians their prime means of reaching national
> audiences and attempting to secure a national following.
> (Mytton, 1983: 117)

The above discussion shows the pattern that laid the foundation for
what has become the largest television network in Africa. If it was
the NTA's policy to have a station in each state (and state governments,
for want of service that was more sympathetic to the requirements
of their constituencies, and for effective control of such channels of
information, could also establish television stations), it was a matter
of time before the mediascape in the nation was dotted with
transmitters. The process was hastened by the quest for distinct
identity that informed the clamouring for creation of more states
out of the existing ones.

To recap on this aspect of the story, the nation at the inception
of television had three regions. By 1967, 12 states had evolved from
these. By 1976 there were 19 states. In 1988 there were 30 states. By
2009, there were 36 states, along with the Federal Capital Territory
(Abuja). With this proliferation of stations, the fear of how to maintain
professional standards waxed stronger.

Nigeria witnessed a number of remarkable achievements during
the second wave of television broadcasting. Riding the crest of this
wave were the audiences who benefited from the proliferation of
stations. Audiences were the justification (if not the reason) for the
service. This was due to the social developments during the period.
The nation had begun to reap the dividends of the free education
programmes that were pursued in the early years of nationhood.
Even among the less well informed, an appreciation for television
had emerged. Television was desirable both as a status symbol and
as a medium of information, if the conditions were right. Contrary
to the belief among media scholars, television in this developing society
was evolving from being the elitist medium that it had been thought
to be.

There were improvements in social conditions at this time. As
part of the Second National Development Plan, social infrastructure

that would facilitate television reception was expanding; more rural settlements were electrified. In any case, the use of petrol generators was gradually gaining ground. The earning capacity of Nigerians had improved since the early 1970s, during the oil boom days and the (Udoji) salary awards following the federal government Salary Review Exercise. The entrepreneurial spirit exhibited in the informal sector meant more people had surplus income. As the informal sector of the economy grew, aspirational lifestyles spread and television began to make its way into Nigerian homes even those where set ownership was less likely. It became the focal point in homes (Esan 1993).

Dividends of the heavy investments made in the earlier days were evident now. The sum of $203.6 million voted for the development of television projects in the Third National Development Plan between 1975 and 1980 seemed to have gone a long way. (Olorunnisola & Akanni 2005: 103). The NTA had upgraded its service to colour. When colour television was introduced, black and white receiver sets became cheaper and more readily available. The local electronics companies that assembled receiver sets deserve some of the credit for this. With these Made-in-Nigeria sets, as well as cast-offs from wealthy relatives or the second-hand market, more people, including those in small towns, could afford to acquire television sets. By 1983-84 a conservative estimate for the NTA reckoned that there were 5 million sets in Nigeria. With an estimated average of six people viewing each set, the NTA boasted 30 million viewers per night. Many were sceptical of these estimates at the time, but anecdotal evidence of the patterns of adoption of television among the urban poor and in rural areas would support this claim. Bamboo poles shooting into the sky from squalid shacks in urban slums, with fluorescent bulbs attached to them, were evidence of the makeshift technology with which television signals were received by the poor. There are tales of people powering their receivers with car batteries. These unconventional means may not facilitate conventional patterns of reception, but they do deliver audiences nonetheless. As black-and-white sets became more affordable, the number of people who had access to television signals in their own homes increased. Those who could afford it were able to view television signals in colour. TV was no longer a novelty.

Another benefit recorded in this era is the development of indigenous expertise. By January 1991, there were 24 production centres and 56 operational transmitters in the NTA. There were 14 state-owned stations as well, all managed by Nigerians. Some had trained on the job locally and abroad, others were educated in institutions of higher learning that had begun to offer relevant courses locally. A proportion was also formally educated abroad. With so many stations around the nation, audiences should have had better access to the stations and more relevant programming. A measure of success was observed in these, as will be discussed in the case studies to follow. Suffice to say at this stage that Nigerian broadcasters found the opportunity to make their mark in this era.

With regard to programming, local broadcasters built on earlier accomplishments such as the coverage of the 1973 2nd All Africa Games, which Nigeria hosted. This had been televised, and thereafter national sports festivals and other cultural festivals were televised. The bigger television moment had been the transmission of the Festival of Black Arts and Culture, which attracted people of African origin from all over the world. The nation had also convened to watch Alex Haley's _Roots_ (1977). Ironically, the mini-series was sponsored by Union (formerly Barclays) Bank as part of its corporate image-building. Barclays' interests in the bank had been taken over as part of Nigeria's sanctions against organisations that persisted in doing business with apartheid South Africa. Thus television, even through entertainment, helped to reflect the nation's foreign policy. This was also evident in the structure of the main newscast of the day. It began with domestic news, followed by news from the African continent, then world news, because Africa is the centre-piece of Nigeria's foreign policy. Attempts were made to consolidate some cooperation with other African nations through the Union of National Radio and Television Organisations (URTNA). There were a few programme exchange schemes but the fact that these did not persist on the schedules suggest that URTNA programmes were not as well received as the Nigerian productions or the Western imports.

With the second wave of television, the days of experimenting with technology for national coverage were gone, and network transmission, with all its challenges, had been established. The NTA was tasked with national coverage, and it delivered. The authority succeeded in convening national audiences on a more regular basis. Audiences converged to watch popular drama programmes. Quarter after quarter, there were locally produced programmes like Peter Igho's *Cock Crow at Dawn* (1980), sponsored by UBA. The schedule also included soaps such as Lola Fani-Kayode's *Mirror in the Sun* (1984) and Amaka Igwe's *Checkmates* (1990). These shows came courtesy of the major advertisers; manufacturers of soap and other consumables, Lever Brothers Nigeria, PZ and Cadbury. The dramas on network like *Behind the Clouds* tended to reflect challenges in contemporary urban society. The mini-series *Mind Bending* (1990), also by Fani-Kayode and sponsored by UBA, was grittier, focusing on drug abuse.

The authority featured some decent comedy in English too— Ken Saro-Wiwa's *Bassey & Company*, for instance. Though it was ridiculed for imitating a British programme that had featured on Nigerian television, *Mind Your Language* (1977-86), another local comedy which audiences associate with the 1980s is *Second Chance*. Some of the old favourites that had thrilled regional audiences—*Samanja, Jagua, Masquerade, Icheoku, Koko Close, Sura de Tailor, Hotel de Jordan*—were also on the network schedule. In this way Nigerians got a glimpse of the programming from other parts of the nation. This formed the basis of the common experience in later years, as can be seen on a number of internet forums, especially those with a Nigerian focus, such as Nairaland.com and Village square. The following section gives an indication of the types of programmes that were remembered by a sample of audiences.

An internet search for "Nigerian Television" will show Nigerians reminiscing about these programmes (whether on private blogs or internet chat forums) and inviting others to share these trips down memory lane. "Salsera" invited people to join in recalling their favourite programmes from the past on Nairaland.com (August 19th 2006). Some even post clips clips of these programmes in their responses. On another discussion thread, started on the 19th of September 2007, "DoubleWahala" posted a link to YouTube featuring NTA commercials from the 80s. "Phantom" initiated a discussion with a message posted on Nairaland.com asking, "Does anyone miss the NTA of the 80's?" on the 29th of March 2009. By the 1st of May 2009, there had been 130 posts. Bloggers recalled a wide variety of programmes. Many were imported but the highest-ranking ones were local. That imported programmes featured so prominently is not surprising because they still formed a large chunk of the programming even though NTA policy was to limit the ratio of local to foreign programmes to 60:40 to preserve the local cultures. In any case, a station like NTA 2 (Lagos) had a different remit that allowed it to have an inverse ratio, with more foreign programmes than local input. In the final analysis, audiences' recollection must be regarded as a function of several factors, including their initial pattern of exposure and the pleasures associated with the experience. From the thread of discussions it was apparent that some of the programmes were those they preferred, while some others they had been compelled to watch—usually by their parents. They had subsequently re-evaluated the value of the experience.

Prominent among the overseas children's programmes were *Sesame Street* (1969-), *Thunderbirds* (1965), *Voltron* (1984), *Super Ted* (1983) and *Dangermouse* (1981). These stood alongside local programmes *Tales by Moonlight, Speakout, Kiddie Vision 1-0-1* in people's memory. Among the factual shows that audiences reminisce over are quiz programmes like *Mastermind, JETS* and *Young Brains*. The network news, especially *Newsline*, was frequently mentioned. This sample of NTA's audience also remembered the 1980s for *Things Fall Apart* (1984), a specially commissioned adaptation of Chinua Achebe's literature classic by the same title, produced by Ralph Adiele. Someone recalled documentaries such as Ali Mazrui's *The Africans: A Triple Heritage*.

The programmes mentioned reflected the age, location and interest of the contributors. *Willi Willi*, an NTA Port Harcourt show, was mentioned, as were programmes on Lagos Television. Some respondents included shows from the 1970s, such as *Adio Family*, that had been on NTV Lagos Channel 10. Though there was mention of Dan Maraya Jos, a Hausa musical, Magana *Ja Rice*, a Hausa drama featured on national network, was not one that many enthused about. Documentaries such as *Food Basket*, which showcased the nation's agricultural resources and promoted federal government activities in encouraging local food sustainability, were not mentioned. Similarly, there was no mention of the *Bala Miller Show* (1983) a major national production developed around a popular Kaduna-based musician.

This was a show on which some of the best talent in the industry had been deployed and it was intended to be suitable for international exchanges (Oni, 1985:14).

Such discussions are subject to the collective memory and preferences of the community of bloggers, which is a mere fraction of the actual audiences. Consequently, their views can only be indicative of the impact of television on this generation. In such trips down memory lane is evidence of how the NTA fostered the sharing of common experiences in the nation. This is the benefit of television, and it is no mean feat that people from different locations and backgrounds can be united in their experiences and share cultural reference points.

Most of the comments about the 1980s were positive. Television in that era was good. Perhaps this should be seen as Nigeria's golden age of television, in spite of the high proportion of foreign programmes. Following are two examples of contributions to the e-forum by Jabbok and RichyBlack.

I miss the NTAs of the 80s real bad. *The Masquerade* [and] *Sesame Street* especially; it comes back hitting my memories like an avalanche. In my view our TV today tends to be efficient, but sterile and dehumanized, revealing a culture that is businesslike and artistically stagnant.
(jabbok, 2009 post 54 nairaland.com/nigeria/topic-255095)

Drama and Comedy:

1. *Cock Crow at Dawn* - I miss Bitrus (Sadiq Daba) and Zamaye (Ene Oloja)
2. *Village Head Master* - Chief Eleyinmi, Amebo, etc.
3. *New Masquerade* - The theme song is beautiful; miss Apena, Chief Zebrudaya, Natti, Jegede Shokoya, Clarus, Gringory and Ovularia
4. *Robin of Sherwood*
5. *Howards' Way*
6. *Matlock*
7. *CHiPs*
8. *Hawaii Five-O*
9. *Jemima Shore Investigates*
10. *CI5 (The Professionals)*
11. *The Avengers*
12. *The Love Boat*
13. *Charlie's Angels*
14. *Doctor Who*
15. *Starsky and Hutch*
16. *The Invisible Man*
17. *Sanduka*
18. *Moonlighting*
19. *Bigfoot and Wild boy* - http://www.70slivekidvid.com/bfwb.htm
20. *'Allo 'Allo!*
21. *Fawlty Towers*
22. *Some Mothers Do 'Ave 'Em* - Frank Spencer is sick!
23. *Mind Your Language*
24. *Randall and Hopkirk* (Deceased)
25. *Good Times*
26. *The Jeffersons*

Educational:

1. *Pronto Brain Match*
2. *Young Brains* - hosted by Peter Okebukola
3. *Mastermind* - anybody remember the name of the host?
4. *Funny Company* - learnt a lot!
5. *3-2-1 Contact* - really inspired me
6. *Wild Wild World of Animals*
7. *Cosmos* - May God bless Carl Sagan!
8. *The Africans: A Triple Heritage* - Thank you Ali Mazrui!
9. *One Northern Summer* - introduced me to the word "tundra".
10. *Kiddie Vision 101*

(Richy Black 2009 post 59 nairaland.com/ nigeria/ topic-255095)

It is noteworthy that the top three programmes listed in both classifications constructed in the above contribution are local; evidence of the appreciation of the effort to produce and source programmes. There were those who felt that the fondness expressed in the memories was merely sentimental; that the past is always preferred to the present, but the overwhelming view in the posts was positive. This may be due to the audience idea of a balanced programming.

The aspersion cast on the current NTA in Jabbock's posting is worth addressing. This and expressed preferences for foreign productions seen in the second example is evidence of the competition faced by national broadcasters with their reliance on internationally syndicated programmes. This pattern had been observed in a UNESCO study a decade earlier! (Nordenstrong & Varis, 1974). The situation occurred in spite of the increase in the total number of hours of locally produced programmes, which, apparently, was unable to match the increase in length of the broadcast day. This was prior to the liberalisation and deregulation of broadcasting that fostered trans-national media operations in the 1990s and beyond (Thussu, 1998, 2004). Clearly there are lessons to be learnt.

Notes

1. There has been an intensification of the situation with the dwindling fortunes of the nation, a trend that began with the Structural Adjustment Programmes and consequent inadequate funding of universities.
2. Consistent with its pursuit for excellence, establishing a Television College was high on the agenda of the NTA. The college, in Jos, was established in 1980, when the NTA was less than five years old. Its first principal, Dr Tom Adaba, was appointed at the same time. The TV College is open to NTA staff and other current or aspiring television broadcasters from within and outside Nigeria.

Chapter 4

Case Studies

In this chapter, the experiences and operations of a sample of state government-owned stations will be considered. These were political necessities, a means of shaking off the hegemonic control evident in the unitary institution that had emerged in the Nigerian Television Authority (NTA). But the question remains: did the state governments fare any better than federal authorities in the service to the people?

Like their predecessors, the stations created by state governments during the Second Republic were established for a host of reasons, but of all these, the task of governance is paramount. Generally, state governors tend to wield more influence in stations located within their domain than they could at NTA headquarters. With the stations owned by their state, there is an understanding that governors expect to have near absolute control.

The real objective of setting up stations is for government to create general awareness of its activities; and to enlighten the populace both in the urban and rural areas. The government wants to be heard, and the federal organ does not give the desired impetus in the state. Imagine! I'm just coming from a swearing-in ceremony of the new commissioners in the state, including the Commissioner for Information, and NTA was not there [with much bewilderment in his tone]. I did not see any NTA cameras there. OGTV [Ogun State Television] can't do that! What's the essence of NTA if it can't be at a ceremony like that? OGTV can't do that! There is no assignment that should have been given greater priority than that. So, to go back to your question, the purpose of a station is to propagate government activities, give it publicity; it should entertain, educate and inform taxpayers.

It can be of use to the generality of the public, for example, publicising their chieftaincy ceremonies. It should cover activities,

policies and reflect philosophy. It could also be used to break age-long cultural barriers to progress. For example, the attitude that we do not number a man's child[ren]. . . needs to be re-addressed if the census is to be a success. These are some of the functions of the station. The enabling edict spells it out.
(Director, Ogun Ministry of Information; interview conducted in 1991)

Incidentally, the civil servant quoted above had once served on the board of OGTV. The short answer to why states establish television stations is the much-bandied cliché—to educate, entertain and inform —which says little of the real motive. Coming from a career civil servant, not a politician, the statement shows the depth of "official" sentiment regarding the role of the medium. There is a flagrant desire for control, with the more noble functions outlined in the early days relegated to afterthought. Enabling decrees, like other such laws, are usually framed in general terms; because they are subject to interpretation, they may convey the impression that television is being employed as prescribed. It takes such in-depth interviews to appreciate how these laws are construed.

In expecting that no other assignment should have greater priority, the director had assumed the editorial judgement of the station without consideration for the exigencies confronting the station, the number of functional cameras, competing assignments and so on. This is indicative of the tensions that characterised television in the Second Republic.

Lagos Television - LTV 8

Established in 1980, LTV was the first state-owned television station set up outside of the NTA. Lagos was a peculiar location for the station because of the prospects accompanying the ability to deliver metropolitan audiences. By this time, Lagos was bubbling with commercial and industrial activities. For good measure the station had the (unofficial) ability to transmit beyond the national border to neighbouring countries in the ECOWAS sub -region. These factors made the station commercially viable, but the issue of governance and cultural identity seemed to be paramount in the consideration of the founders. There had been a station in Lagos since 1962 (NBC-TV, later NTA Channel 10) but the indigenous people of the state had not been well catered for. They had been sacrificed at the altar of national consideration. Civilian politics offered people the opportunity to pursue their dreams; it was time for the people of Lagos to have their own TV. The politics of its establishment has already been discussed. According to a manager, Programmes, of Lagos State Television,

> LTV is a state-owned television, so we consider the populace, the special audience of Lagos state. Lagos is a dual capital, federal and state, so we have diplomats, foreign missions, Lagos state indigenes [including] Aworis, the Ijebus, the Eguns and the migrants from other hinterland who are in search of the golden fleece with all the industrial locations within Lagos.
>
> The majority of the indigenes are Yoruba-speaking. This in essence guides what we beam out. The state TV belongs to them; they are the taxpayers financing LTV.
>
> It is not unusual that we have a Channel 10 [NTA] approach to programming but it is also like Channel 7, which is geared towards the grassroots.
>
> (Manager, Programmes, LTV Interview conducted in 1991)

This last remark shows that the station regarded its orientation as a cross between that found in two Lagos-based NTA stations, acknowledging the city of Lagos, its cosmopolitan audiences, as well as indigenes of other areas within the state. In fact LTV may have taken a cue from Channel 10 but it set the pace for Channel 7, which came after it. LTV prides itself on being a people-oriented station. This had always been the plan. A remark in an interview with the founding governor, Alhaji Lateef Jakande, to mark the station's 27th anniversary shows how idealistic it was meant to be. Jakande, himself a journalist, tried to make a distinction between government, people and state. Many may regard this as an esoteric distinction.

> LTV was not established for government publicity. It was to give a voice to the people of the state, and to project the state. LTV over the years has lived up immensely to this vision.'"
>
> (LTV-Next Nov-Dec 2007 p10)

Over the years the station has adopted different slogans to affirm its brand. At its inception, its slogan was "8 Alive" to celebrate the fact that it was in operation despite attempts to crush it. It had been shunted to Channel 8 as part of the resolution of the dispute with the federal authorities over Channel 5. Its signals were also on the UHF band, which was where other state stations were confined. LTV is just an arm of the Lagos State Broadcasting Service. It shares its premises with the radio service—Radio Lagos. Staff have been known to move between the services, as have certain programme ideas, yet operationally there is little overlap between the services as there had been in earlier broadcasting efforts. Yet LTV is associated with the corporation's Yoruba sign-off *"Tiwa 'n Tiwa Mititi"* [meaning "We're indigenous and we rock!"]. With this philosophy it identifies with the people of the state. Its mission of creating for them a sense of ownership of the station and a sense of belonging in the state is thus apparent.

Its core staff at the inception of its service had come largely from the NTA and the Federal Radio Corporation of Nigeria (FRCN). They were drawn from different cadres but generally these were people with years of broadcasting experience, who could be assumed to be familiar with the professional cultures in the industry. Though there were grand plans for this station, and Lagos state had the financial muscle to realise these, there was intense opposition. Within five years of starting operations, coming after the dispute over airwaves, its premises were gutted by fire. Regardless of such setbacks, LTV has bounced back.

The organisational structure at LTV is fairly typical of other stations regarding the delegation of duties. It has eight separate directorates: News, Current Affairs, Administration, Finance, Programmes, Programmes Services, Marketing and Engineering. However, unlike other stations, it has been headed by a permanent secretary since 2001, when chief executive Jimi Odumosu (appointed in 2000) was upgraded to permanent secretary. This distinction is crucial. It makes no bones about the place of the station within the governance system in Lagos state. Broadcasters are in no doubt about being a part of the civil service machinery. Not for them the illusions of being an independent parastatal, though it has been argued that this need not undermine the service to the people.

> It is a government-owned station. The mission of LTV therefore has been to provide the government an opportunity to reach the people and provide feedback for both the government and the public. Because of that, we have found ourselves in the unique challenge of balancing both sides, being socially responsible and believable.
> (Lekan Ogunabawo, permanent secretary, LTV interview conducted July 2008)

This honest appraisal of the relationship between the government as proprietor and management as professionals could only be adopted in the light of experience. As stations and government officials have matured, the orientation to the use and management of the medium has also evolved. Government officials are starting to learn that the medium will be a more effective tool if they let it remain impartial. For example, unlike some state-owned stations, which heed party resolutions to embargo advertisements of opposition candidates, in the 2008 elections LTV made a convincing case for giving equal access to all parties during those elections. This attempt at impartial coverage helps to generate the trust of the public and the distinction between a government-owned station and a government megaphone.

> [It is] government-owned but not [a] government megaphone. LTV manages to give a voice to various audience groups. We are the government mouthpiece in a positive sense. We are critical of government and we make suggestions.
>
> The LTV gives a voice to the Lagos state government. By watching the station the audience knows what the official government position is. The government puts out notices on the station and the audience can trust this to be the government's position, even if they disagree. Information on other stations may be unsubstantiated, mere speculations."
> (Permanent secretary, LTV interview conducted July 2008)

This may suggest that the station is little more than a noticeboard, as may have been the case under the military and more autocratic politicians, but LTV has always tried to balance its duties in disseminating public information with the provision of a range of entertainment programmes that have an appeal to the more culturally diverse audiences of Lagos state. The indigenous audiences are mainly Yoruba-speaking, but the station is still responsible for migrants from other parts of the nation who have settled in Lagos, and its programming reflects this. This culture was established by the man who conceived the orientation of the station, and became the first coordinator of programmes, Taiwo Alimi. His focus on grassroots audiences was sanctioned by Governor Jakande, who believed that the station was to serve the range of taxpayers in the state.

The Programmes department reflects this emphasis, as it is split into two main units, Entertainment and Enlightenment. There is no pretence of engaging in educational service. Entertainment at LTV consists of the Drama unit, and Music and Variety. These provide opportunities for learning, though not through a formal educational structure. Women and Children issues were catered for under Enlightenment, as were Public Affairs. Again, these are avenues for informal education. Other than these, it appears that Chief Awolowo's initial conception of television as a teacher had not been formally adopted in the more recent incarnation of the project. The shift is as much a reflection of changes in the broader context as it is the personal philosophy of those with responsibilities for interpreting and executing policy

> Entertainment is mainly for the enjoyment of all [and is] primarily for relaxation. In productions that are meant for enlightenment – for example, vox pop which goes round town sampling people's opinion, or the Women and Children's programmes – there is an attempt at informal education.
>
> My personal philosophy is that if you want information, you might as well pick an encyclopaedia. But people want to be entertained and [to] learn. So if you want people to watch, you will make learning fun; envelope the package in fun.
>
> (Manager of Programmes, LTV Interview conducted in 1991)

This marks a generational shift away from the high premium placed on (formal) education. Barely 30 years after the Western regional government initiative that introduced television as a teacher, the nation was already reaping the dividends of free education programmes. Opportunities for formal education had also increased. A higher proportion of the population was literate and could explore other avenues for gaining knowledge and information. These may account for the apparent change in tactics employed in programming. In any case, the wider experience of formal educational broadcasting had shown how involving such a mission needed to be.

LTV has a good track record in providing entertainment, which helped it create a rapport with the audience. Its weekend service, Lagos Weekend Television, was the first 24-hour television service in Nigeria transmitting from Friday to Monday morning. This feat is credited to Alimi and his team (Jakande, 2007). This gave it ample time to serve various interests in the market: those who loved it for its local dramas; those who desired the latest musical videos; and those who wanted the foreign films. It was a precursor for the 24/7 service that the station now offers. In financial terms, that decision may not be viable but the station persists, to sustain viewer loyalty. The 24/7 service builds up goodwill and is pursued in the belief that the audience base must be maintained, not pushed away. People may be lazy in switching loyalties in spite of the ease of the remote control, but this may only last as long as they are kept happy.

Providing avenues for entertaining may also help to achieve the vision of broadcasting for the people. Having brought them to the channel, the station is able to disseminate information on behalf of the government and the corporate world. It is regarded as an *"Mgbati"* [ie, Yoruba-speaking] station though it has programmes in Ogu (Egun), Awori, Igbo, Hausa and English in recognition of the diverse ethnic base of its audiences.

In many ways the 1980s were the decade of "austerity measures" yet that was when this station was born. Its travails were compounded by the fire that gutted its premises in 1985 destroying its prized facilities and much more. The station lost equipment, recordings of productions, including drama, rushes from shoots, news reports, and vital evidence from the coverage of elections (and alleged rigging). The fire was suspected to be an act of sabotage. What was worse was the military takeover of government from civilians again.

The military authorities had little need for a distinctive service, finances were tight and the station's struggle to stay alive began again. Yet it remained defiant. The early staff were passionate about the service, doing whatever it took to keep the revenue from sales flowing in. LTV was noted for its Father Xmas Funfair, a carnival-like celebration held on its grounds. This idea, credited to one of its producers, Kate Omitosin, was brought to life in December 1984, and has been an annual event ever since. At these, a new twist was brought to audience requests so popular on radio. Audiences could send video messages to their loved ones live from the station. The Father Xmas Fair was an example of how LTV democratised access to the media, as schoolchildren who (because of their lack of social capital) might never have had the opportunity to participate in regular television programmes, had the opportunity to be on air, sending season's greetings. There were also commercial benefits for the station and the participants. There were opportunities for publicity for businesses. Many exhibitors, along with food and drinks vendors, did brisk business at such events, which made the nominal charges for participation worthwhile. Coverage of the fair was also a source of programming. In the dark days, other types of fairs and sponsored programmes, particularly religious programmes, thrived on air. In the face of sustained under-funding and neglect of staff welfare, spirits flagged and perhaps visions waned.

The 1990s were challenging times at the station as in the nation at large. The deregulation of broadcasting begun under the tenure of General Ibarahim Babaginda would change the media landscape. Existing stations had to respond to the competition. When private stations started operations, LTV was flanked by competitors that muscled in on its patch. Lagos was a proven profitable patch, the LTV programming format was one to be cloned, and the stations lapses were brought to light by those who sought to improve on its performance as its fortune dwindled.

In this period LTV was still under different military governors who felt no compulsion to equip the station and keep it ahead of the game. Lagos state had varied experiences under these military administrators. Apart from the varied levels of dynamism they exhibited, it was hit by wider political and administrative developments. The movement of the Federal Civil Service out of Lagos (to Abuja, the Federal Capital Territory) left Lagos depleted of the funding and attention that the federal presence guaranteed. This took its toll on the people and government of Lagos state as well as their television. By the end of the decade LTV needed serious capital investment. The premises were rundown, vehicles in the fleet were grounded, morale was low. Many staff felt curtailed and vulnerable, especially in the aftermath of the annulled June 12th 1993 Elections. As party lines of old were maintained, the affiliations and sympathies in Lagos were largely in favour of Chief MKO Abiola, whose mandate had been denied by this annulment. It was also a slap in the face for the electorate who had given the mandate. A new era of press activism had been ushered in with the annulment. Broadcasters had to be more cautious, and television was less flexible than radio, which had been adopted as part of the struggle for democracy. Radio Kudirat was a pirate radio station that operated at this time. There was no television equivalent. This must have heightened the sense of frustration, especially during the regime of the late General Sani Abacha, whose repression of pro-democracy activities earned Nigeria notoriety in the international community and expulsion from the Commonwealth. Yet the team carried on regardless. Their human rights agitation found expression in support for civil society initiatives for the vulnerable members of society—women, children and the poor. Yet the situation had degenerated so much that the station had to operate without basic facilities. On occasions VHS cameras, microphones, editing facilities had to be hired from commercial operators, in order for programmes to be made.

By 1999, when the nation returned to civil rule, the station had to be re-engineered and repositioned in the market. The onus fell on the civilian administration of Asiwaju Bola Tinubu to inject the required funds. The staff team also had to deliver. The process of recapitalisation began in 1999 because politicians appreciated the merits of the station. The state Commissioner for Information had been one of the pioneering staff at Lagos State Broadcasting Corporation (LSBC). In any case, politicians need votes from the people and therefore needed the station to promote their activities and themselves. They also had to keep the media on side; staff at the station needed to be encouraged. The station seized the opportunity to catapult itself into the 21st century; to respond to contemporary global influences, not just seeking to recapture past glory.

New transmitters have now been acquired. Usually the station had just one, at most two. LTV now has four transmitters (50kw, 40kw, 20kw and a stand-by 2kw transmitter). The station operates a 24-hour service and it cannot afford to be down. It has also acquired two 500-KVA generators to guarantee uninterrupted power supply.

The cost of maintaining these and generating electricity is phenomenal. It constitutes part of the constant challenge of running a television station.

> Staying in business in spite of daily challenges is in itself a joy. For example, in a month, the station pays up to one million naira to [Power Holding Corporation of Nigeria, PHCN] for undelivered service. On top of that it spends seven million naira on diesel to be on air.
>
> Whereas if PHCN charged three million naira and power supply was stable, that would relieve the station of the burden of buying, running and maintaining the generators, then hiring the personnel to man them.
>
> (Lekan Ogunbanwo, permanent secretary, Lagos State Television Interview conducted July 2008)

In addition, the station has invested in the tools of its trade—digital cameras, lights, microphones, non-linear editing suites. The entire premises have been renovated, uplifting the spirit of the staff. Many, including news reporters, editors, drivers, have been supported to go for training courses; a number have benefited from a loan programme under which they have purchased their homes at a government housing scheme. These are benefits of being civil servants in Lagos state. Management has taken up the responsibility of looking after staff welfare and creating a strong family bond.

Provision of staff welfare has clearly been part of the strategy employed to revive spirits in the station. This includes creating a conducive environment. Building a staff clinic, an improved canteen and transit accommodation for personnel who work odd shifts are examples of gestures which help to make staff feel appreciated. The station is blessed with private entrepreneurs who are able run profitable ventures on its premises. The newsroom was also given a facelift between 2005 and 2007.

The renovation of the premises is part of the station's financial strategy; the station is again able to offer halls and spaces in its premises for hire. This is an indirect means of keeping in touch with its audiences. More formal corporate affairs efforts help management communicate with its various audiences: there is an in-house journal (LTV Next); monthly meetings with staff; marketing officers making pitches to clients and helping them make the most of improvements in the station. They can do this from a point of conviction. An observational study of the station (July 2008) shows that it has regained its energy and passion. Judging by their contributions to the in-house journal, members of staff feel ready to face the world. Management is also leading by example, with the chief executive still on news-reading duties.

Since June 2008, LTV has been on Digital Satellite Television (DSTV). Channel 129 was thus able to legitimately reach audiences on the African continent. The attempt at delivering via the internet (on the JUMP TV platform) has not been as successful but the opportunity has been recognised and is being explored.

Besides government funding, LTV's management does try to be creative in the business models it employs. Securing the DSTV deal was a result of management's persistence and bold business propositions. There was a desire to reach its audiences abroad, yet relying on the satellite transmission through DOMSAT would have been too expensive. By offering DSTV the opportunity to share the site and height advantage of LTV's transmission masts, LTV management was able to get a cheaper deal to fulfil its dream. The station LTV got a platform for transmission in return.

Management is delighted that it is being watched locally and in locations as far away as Sokoto (north-west Nigeria), Côte d'Ivoire and Ghana. Managers are quick to point out that this audience reach is not based on scientific audience research, but on calls that come in during audience participatory programmes; live shows since the station has been on DSTV indicate that the station is being watched in those places.

"Even when the calls are critical [of the station] the comments are indicative of the benchmarks that the people have set expectations . . . of the station. This is heartwarming, as it shows that people care."
(Lekan Ogunbanwo, permanent secretary, Lagos State Television Interview conducted July 2008)

Such expectations can only help the station to improve the service. There are collaborations with some of the local Yoruba artistes to produce videos which are later serialised on television. LTV has sponsored some movies that it first marketed before transmitting. It also conceived the idea of providing production services to people who have ideas but require technical know-how to translate their ideas into programmes.

The management is aware of the limitations of the content provided on its station. Though this is attributed to the skills deficiency of fresh graduates, there is evidence that poor recruitment practices still remain. When institutions lack the equipment for training, they turn out graduates without prerequisite skills. The skilled professionals in the industry fix their charges so high that stations cannot afford to hire them. Instead stations end up with people who are passionate about television and are willing to learn on the job to overcome their skills deficiency.

Still, LTV has a niche in the market. It is the pride of its proprietors, the government and people of Lagos state. It does not lack for endorsement and commendations from those who envisioned it.

> I am happy to say that LTV has lived up to my dream. The station has achieved all I set out to do. It served as a pacesetter to other state-owned stations, put up so much brilliance that keeps it on . . . it worked because it became the life of the people. I knew that LTV [would] compete with foreign stations.
> (Alhaji Lateef Jakande, ex-governor of Lagos state interview with LTV-Next, Nov-Dec. 2007 pg. 10)

Fig 4.1: LTV Transmitter Controls

Fig. 4.2: Refurbished Master Control Room LTV

Fig. 4.3: Generating Electricity

Fig 4.4: Newscaster on Duty - Lekan Ogunbanwo, LTV Permanent Secretary

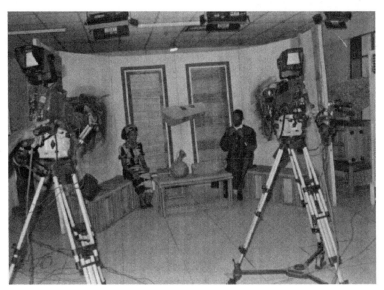

Fig 4.5: Igbo Presenters on Set

Fig 4.6: Sponsored Promotional Programme at LTV premises
Images Courtesy of Lagos Television

Attestations from a happy workforce at Lagos Television

"This is a clarion call to committed and dedicated staff of New LTV . . . Here is a new era of epoch-making transformation in the annals of Lagos Television. The dramatic changes and reforms going on in this organisation can be perceived even by a blind man. . . We need to take cognisance of the fact that courtesy of the new LTV management have had standard training courses, sophisticated broadcast equipment and jumbo sized welfare. How do we justify these financial investments of LTV?"
—Anifowoshe Abiodun, Programmes Directorate in *Workers Arise* Correspondence Section LTV Next July 2007 Pg 4

"I see the organisation as an embodiment of my personal Philosophy of focused growth and sustained development . ."
—Amos Adeyinka Adeyanju Directorate of Marketing In Our People Section LTV Next Nov/Dec 2007

"No doubt, we've added value to television broadcasting in Nigeria... We've

unlocked …imagination on how a public service information channel can be successful…"
—Solomon Gbadebo in Our People Section LTV Next July 2008 pg 14

"To God be the glory. One can really walk tall and say;"I belong to LTV" at public functions
—Omololu Rosanwo

"I am not a bit surprised. I know you can do much more"
—Afolabi Ogunjobi (Bayowa Films) Comments in Finally Out of Darkness LTV Next
April 2007

"… The glory of Lagos Television in the eighties is on the way back…"
—Adeyemi Olowosagba, Corporate Affairs Unit in front page article Two Years of Greatness Celebrated LTV Next April 2007

Ogun State Television (OGTV) - Gateway Television (GTV)

The story of television in Ogun State (Gateway Television, or GTV, formerly OGTV) is similar to that of the other state stations. However, it had to contend with situations that were peculiar to itself. To begin with, it was operating in a state which was right at its infancy, having been created in 1979. It was one of the states carved out of the original Western region (later state). Although technically Lagos and Oyo states were also created in the same year, those locations had had the administrative machinery of state governments in their state capitals, Lagos and Ibadan. These were the locations that had inherited the early infrastructure to support television, as well as a wealth of staff experience. They also had better established levels of commerce and other features of cosmopolitan life that advertisers desired. These did not exist on the same scale in Abeokuta, the capital of Ogun state and the location of OGTV. That was why it was important for OGTV's reach to extend to Lagos and Ibadan. But there was a problem from the outset.

Like Lagos and other state governments, Ogun state was caught in the 1981 tangle of policy regarding frequency allocation. The state governor, Chief Olabisi Onabanjo, had a different approach to his colleague in Lagos. He was going to make history, by establishing the first television on UHF in Nigeria while other states were busy protesting. The station was to cover the entire state, not just the capital, and it was to be commercially viable, hence the need to reach beyond the state. It appears that his defiance was being expressed in different ways. The people of Ogun were required to prove their mettle.

The OGTV example shows that state government television stations were largely parochial. From the beginning the planning team consisted of Ogun State indigenes who had varying shades of expertise. With the challenges that UHF transmission posed, much depended on the guidance received from the all-Nigerian technical team. Their first task was to locate the right site for the station. Another challenge was to minimise the costs involved. Because of the need to reduce obstacles in the way of the signals, it was important that the transmitter be located on a very high elevation, thus reducing the need to amplify signals. The terrain had to be able to support the weight of the structure for the mast and it had to be accessible. If it was not close to existing infrastructure, roads, electricity, telephone lines, water and other expensive essentials would be needed, adding to the cost. For these reasons, OGTV was situated on the outskirts of Abeokuta in a place close to an existing road but which lacked any other infrastructure. (Eleshin , 2006: 121 – 136)

To reduce costs further, the transmitters and studio facilities were to be on the same location. Previous practice had been to have transmitters (and, by extension, the engineers and technical staff who maintained them) in the middle of nowhere, while studios were closer to urban centres. On this occasion the engineers succeeded in taking the entire team to the back of beyond. They had a good excuse—it would cut costs! Another clever move was to make OGTV a residential station, at least for key operational staff. With these workers living on site, it was easier to respond to emergencies and monitor operations. The frugal people of Ogun state had risen to the challenge of UHF transmission. They established a station that they were proud of. The following is an excerpt of the speech made by the first executive governor of Ogun state for the formal launch of Ogun State Television at its permanent site in May 1982.

The station we are commissioning today is a shining example of international cooperation. It comprises three studios, technical services units, transmitter hall, office units and massive balcony facilities for outdoor recordings, central air-condition facilities, among others. The station is an integral part of the Ogun State Television Village, Nigeria's pioneer television village, which is also being launched here and now. The first phase of the television village consisting of staff quarters has been completed. Other phases to be embarked upon include additional quarters for senior and junior staff, provision of supermarket and recreational facilities, and the establishment of wide-ranging urban and rural settings for filming purposes. So the whole complex has cost us 6.5 million naira.

We are today one of the four states to own its own television station. The others are Lagos, Anambra and Bendel; but today we become the first that has a station which covers its entire state, and broadcasts beyond. The station has cut an enviable image for itself. (Olabisi Onabanjo, executive governor of Ogun state, 1979-1983; founder of OGTV (Onabanjo, 2006: 8 – 9))

At inception the station was operating out of a temporary location. The foundation stone for the permanent site described above was laid in June 1981 and the station had moved in by May 1982. Because there was no public supply of electricity, water and telephone services, the station relied on private electricity generation for all its operation for several months. This was a very expensive, but it was so reliable that even after being connected to the less reliable public power supply, the station preferred to rely on its private generation of power for transmissions.

It had always been designed to be a commercial station, and no time was wasted in setting about this task. But it was hastened along that path by the financial predicaments of the various governments. OGTV was cut loose from the financial ties that might have cushioned its operations. It became the first TV station required to break even in its operations and meet its recurrent expenditure. It had done so by August 1984, not even three years after beginning, before it could really try and test the patterns that it would evolve (Adebimpe, 2006: 236). Unlike earlier broadcast management models, there was little synergy between OGTV and its sister radio station. The successful state-owned radio service Ogun State Broadcasting Corporation (also known as Ogun Radio) was kept independent of the television service;

indeed, they were miles apart. This was another score on which Ogun Television was handicapped from the start, yet such challenges merely propelled the team to be even more resourceful.

The history of the station is a glaring illustration of how television was a result of the determination of the political elite. Certain decisions taken in the pursuit of such a vision could be described as whimsical rather than a response to the exigencies of the state. The decision to commence transmission on December 25th 1981 was said to be informed by the desire of the governor to present the station as a Christmas gift from the government to the people of the state. That decision was supposed to have been influenced by the frequency allocated to the station. OGTV was on Channel 25 UHF.

> "We started with the most skeletal staff [I've known] in my life. I was the founding officer in charge of programmes. We moved to the present complex after it was commissioned on the 13th of May 1982.
>
> Transmission was from hand to mouth, for a period of about four and half hours a day. In spite of the fact there was not much else, we had ambition. We used OB [outside broadcast] vans for transmission . . . Pioneering staff of this station had not had experience of television. The job was cumbersome and rewarding. We hadn't any precedence in terms of patterns to follow. We evolved our own patterns ourselves."
>
> (Assistant general manager, OGTV Interview conducted in 1991)

This officer confessed to his lack of television experience before working at OGTV. Other members of the team had varied levels of experience, especially as artistes at other (NTA) stations. This he considered to be an advantage in the sense that the team was thus were not bogged down in industry cultures. They were free to experiment and be innovative and able to introduce some freshness into the service. The station was of the view that they were meeting gaps that existed in the market.

AGM: Something else we did was to introduce what we called Breakfast Television at the weekends. As I said, I like classical things. This was introduced during the Easter weekend in April 1982.

Researcher: Why?

AGM: I don't know. Maybe it was my romantic nature. We
             wanted to do something unusual. We had ambition.
             We did not have much else but we soon became a force
             to be reckoned with. We transmitted movies, excellent
             musicals and a few solid local productions.

The station had clearly stretched its staff and resources to fulfil
ambitions. Unlike the patriarchal station, WNTV, attention to the
purpose of television was not as carefully thought out. Some practices
at this station (as at many others) were questionable professionally,
yet these were desperate times. The fact that the station was on the
outskirts of Abeokuta, beyond easy reach of advertisers, since there
were no telephones, only compounded matters. In those days, stations
did what they had to do to achieve what in the end may be described
as innovative and popular; the two keys that they identified as necessary
for successful commercial practice (Adebimpe, 2006:238). The
stations needed to look beyond the traditional channels of revenues
—advertising agency sales, announcements and social diaries. New
twists were introduced in programme sponsorship so there were
full, joint and segmented sponsorships. Production facilities were
commercialised, so studio and equipment hire was introduced. The
station was involved in promotions, film shows, concerts and the
controversial commercialisation of news. At the time, there was less
concern for standards, but more consideration for audience pleasures
and the generation of staff salaries. Yet the station appeared to have
been a hit with the target audience, as the quote below shows.

> Because Ogun state is a gateway state, opening into the Republic of
> Benin on one side and to Lagos and the outside world as well, we
> had dubbed the station The Gateway Telly. But people, of their
> own volition crystallised the name The People's Telly and that is how
> we came about the name. We decided to take it on, since the people
> themselves had re-christened us in appreciation of our efforts.
>   The essence of programming at this station was 'to give the
> people what they want', having seen NTA for so long and how
> frustrating and boring that had become.
> (Assistant general manager OGTV Interviewed in 1991)

This came at a high cost. Staff cannot forget those days of uncertainty, not knowing if salaries would be paid or if artistes' contract fees would be released. The station was either not paying or was delaying payments; in any case government subsidy was only responsible for capital expenditure such as major repairs, rather than running costs like salaries and wages. These practices changed the face of broadcasting, as noted in the view below from a member of the academic community.

> During the era of news by barter, news was dominated by pseudo events—annual general meetings of corporate organisations, obituaries and burial ceremonies, conferment of chieftaincy titles and other mundane routine activities of government.
>
> Another eagerly seized-upon option was the sale of airtime to independent producers and corporate organisations. The practice more or less saw to the end of the quarterly programme planning and scheduling in many broadcasting stations. Programme broadcast schedules became subject to change at any time, so long as a willing buyer showed up. A good number of radio and TV stations stopped publishing their programme schedules for members of the public to access. Programme transmission became so flexible, unstable and subject to constant and instant change. Listeners and viewers were not sure of the time allotted to any programme or whether a particular programme would last a quarter.
>
> (Oso, 2006: 265)

Though these practices were widespread, particularly at state government television stations, as observed in an observational study conducted in 1991, OGTV proved to be the most ruthlessly commercial of the stations in the survey. (Esan 1993) Commercialised stations featured productions with a questionable mix of claims and values by traditional and spiritual healers, with little effort made to verify those claims. Religion (Christian and Islamic evangelists) became a brand on OGTV, and personal celebrations of the privileged tended to dominate the screen. The assistant general manager (Production Services) was most unapologetic for this trend even when he found some of the programmes distasteful.

I have looked at the socio-cultural setting of the station and I have designed strategies that will help us to make money and survive within it. The Yoruba man is fun-loving, he likes parties, titles and publicity ... We have problems with funding ... it is not peculiar to television; it's all over. Funds are very hard to come by and costs are going up. Costs are going up and funds are not coming in. There's a rethinking in terms of what is to be done, and how to do them, and this is where ingenuity and creativity really come to play. You use what you have to get what you want ... It's tough, there's no doubt, it's tough, but we keep managing. We keep adjusting to the realities of the situation.

(Assistant general manager, OGTV Interviewed in 1991)

Such adjustments have also meant that the station is prepared to broadcast programmes from bodies or individuals who can help procure much-needed spare parts through contacts abroad, even if they cannot pay with cash in Nigeria.

The flexibility of this trade by barter and the "rethinking" are examples of the ingenuity that individuals in the organisation took pride in. They illustrate the impact of commercial motivation in the running of a station. Some of the ideas, such as the social diary, were no novelty. The same idea had been used (perhaps more modestly, with fewer minutes of coverage per event) on WNTV. Some of the less conventional innovations were an embarrassment in professional quarters, but these were products of the time. They were born out of necessity, as the quote above indicates. This was the period when the Nigerian currency was taking a dip in the foreign exchange market, which affected buying power. Spending on foreign programmes was carefully managed, with priority given to movies that would draw in the audiences. Collaborations with organisations like the Christian Broadcasting Network and indigenous theatre groups yielded programmes. To augment these, the station also relied on the cultural service of certain foreign missions in Nigeria. These were austere times for individuals, governments and their organisations. Certain policies might otherwise have struggled to find approval within a television organisation. Adopting them was possible only because of the commitment of very senior officers and the desperate situation. This was a watershed in the quality of practices on television stations.

Table 4.1: TV Weekly Output Pattern Total Minutes by Genre - 1991 data

	NTA Ibadan	NTA Abeo-kuta	NTA 10 Lagos	NTA 7 Ikeja	OGTV	LTV/ LWT	BCOS
News	935	1025	685	585	1105	1050	1030
Drama	450	630	525	420	630	545	295
Education	150	180	150	90	60	0	90
Children	270	380	570	300	165	210	125
Religious	115	60	110	60	480	510	360
Youth	60	0	60	60	30	0	0
Health	0	60	0	0	30	90	0
Metaphysics	0	0	30	0	210	210	0
Women	30	30	30	60	60	60	60
Talk	390	315	225	390	470	105	290
Enlightenment	0	150	30	0	60	185	30
Government	180	150	120	180	0	60	85
Entertainment	90	90	190	90	0	60	85
Musicals	180	105	320	90	60	320	230
Movies	120	0	165	0	630	180	870
Magazine	0	0	0	0	90	105	180
Cookery	30	0	30	30	0	30	30
Documentary	60	60	150	60	0	0	60
Sports	325	210	270	300	240	210	345
Promos	60	60	0	105	265	0	215
Unclassified	90	0	0	0	0	30	30

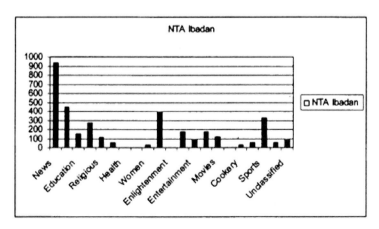

Fig 4.6: NTA Ibadan Output Graph 1991

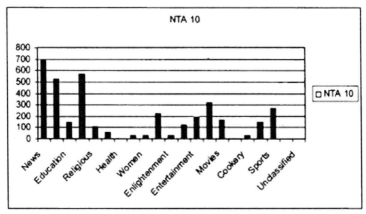

Fig 4.7: NTA 10 Output Graph 1991

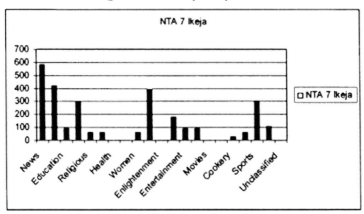

Fig 4.8: NTA 7 Output Graph 1991

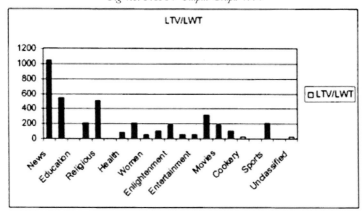

Fig 4.9: LTV/LWT Output Graph 1991

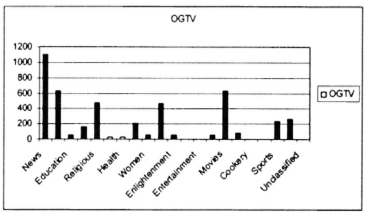

Fig 4.10: OGTV Output Graph 1991

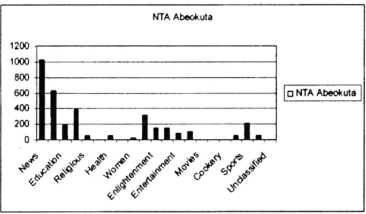

Fig 4.11: NTA Abeokuta Output Graph 1991

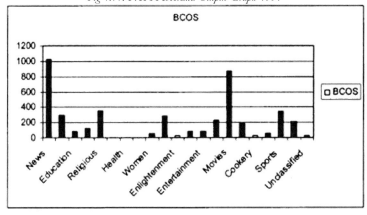

Fig 4.12: BCOS Output Graph 1991

A critical lesson from the experience of OGTV in the 1980s can only be fully appreciated in the context of the organisational structure and staffing structure. As in most television stations, a general manager is at the helm of affairs. He is the link between the station and the board of governors. The board was a politically appointed group of people who set the station's policy. It was the ultimate in the hierarchy of power at the station, outside of the state governor. The board represented the interests of the people of the state, as it guided the management of the station. The general manager, with his team of directors, translated and managed the implementation of these policies.

The station was supposed to have five directors; one each for Engineering, Programmes, News and Current Affairs, Administration and Finance, as well as Commercial Services, but the appointment for Commercial Services, was initially held up for several years. Clearly this directorate could not have been regarded as central to the existence of the station. Borisade's discussion of the orientations to management of television stations confirms the claims of superiority by certain departments in the hierarchy of television. He spoke of two dominant schools of thought within television management in Nigeria, even in the premier station (WNTV Ibadan), where officers from Administration, Finance and Marketing were regarded as less relevant to the core business of television than the heads of Engineering and Programmes.

> And to show how powerful the two were, the director of Engineering (chief engineer) and the director of Programmes were placed on higher salary levels of £2,780 per annum while the director of Administration, director of finance, director of Commercial Services and director of News were on £2,500 per annum. When the post of general manager of the station became vacant in 1973, only two candidates—the director of Engineering and the director of Programmes—were interviewed for the post.
> (Borisade, 2006: 142)

The relevance of those other departments became immediately apparent in OGTV, as had been the case in the wider industry. However, when the station's operational priorities changed, its organisational structure had to be adapted. The organisation was collapsed under two key operational umbrellas with the introduction of the posts of assistant general managers (AGM). There was one for Engineering Services and one for Production Services. News and Current Affairs, Programmes and Commercial Services were under AGM Production Services.

Engineering Services was responsible for all maintenance issues; transmitters, studio equipment, buildings and transport. Production Facilities comprised studios, library and graphics, including stores and editing. It was responsible for purchasing equipment spares, programmes and software for the facilities (on the advice of the engineers). Production Services was also responsible for studio operations teams. The Presentation Unit was another unit within the department. This was the unit for the announcers, presenters, moderators and other in-house on-air talents. Each department was headed by directors, just like the less esteemed Commercial Services department, and they were all to report to the AGM. When that post was to be filled, it was the director of Commercial Services who was promoted. This appointment crystallised the elevated status of Commercial Services. The hierarchy of other staff in the organisation is typical of other television organisations, as will be discussed in a later chapter.

At OGTV, officers were moved from Programmes, the spending department, to Commercial,, the earning department. The Commercial department had hitherto been a dead end at this organisation, but with the changes described, it was reinforced and exalted to being the most influential and most rewarding department. This reflected the priorities of the station. Staff with productive records had to be drafted into the department when the station was required to be more commercially aggressive. Performance was rewarded with promotions. The speedy career progression in the revitalised Commercial department was, however, the cause of resentment among those staff who felt better qualified and more deserving. The experience shows how important every arm of the team is in the television business. Had initial recruitment to the departments been mindful of this, the station might have averted such aggravation. Clearly the impact on staff morale of such organisational restructuring should be considered in planning and decision-making. While it may have been too late for some, the situation appears to have been rectified, as a distinct post, AGM, Marketing and Sales, had been created by 2008.

Given the new demands of the organisation, there were more changes to the schedule of duties. Each member of staff, regardless of previous designation, was considered a salesperson. They were induced by the commissions for the sales they made. This arrangement was in addition to the corps of employed sales officers and the sales agents who worked for the station.

Fig 4.13: Typical Layout & Hierarchy in a TV house

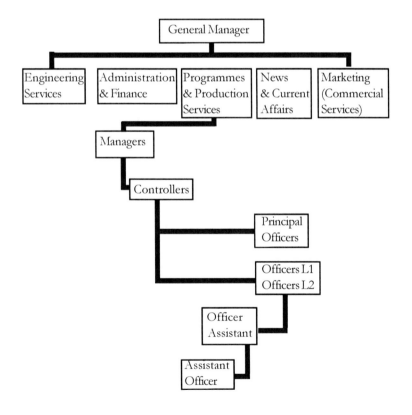

Irrespective of how this system was justified, it was unpopular among certain staff. Some producers, for instance, felt their creativity was being stifled, reckoning that the odds were especially stacked against them when they focused on rural audience segments, a group not readily attractive to advertisers. In this way, the social responsibility objectives of the station were being threatened by its quest for economic survival.

The system had also discouraged some "hardworking creative staff" who lacked contacts and could not raise external financial backing for their programme ideas. In a system where sponsorship was the paramount consideration, productions that even the staff of the stations acknowledged as mediocre and unprofessional made their way on to the screen Even at management level, personal sentiments against some of those sponsored programmes ran high, but as long as they brought in much-needed funds, they were tolerated, even welcome.

On the other hand, some programmes of considerably better quality were never developed beyond proposal stage for lack of sponsorship. Invariably such productions failed to attract sponsorship because they would require higher budgets (not to mention expertise). Sponsors had a choice of other, possibly more central stations, and increasingly found independent producers an attractive and economical option. The corps of independent producers was swelled by the disgruntled and frustrated staff who quit the various stations, having found independent productions to be more profitable.

	N (Naira)	%
NTA-2 Channel 5 (Lagos)	1, 300, 000	33.29
NTA Ikeja (Lagos)	170, 000	9.35
LTV (Lagos)	450, 000	11.52
BCOS (Ibadan)	240, 000	6.14
NTA-10 (Lagos)	110, 000	2.81
OGTV (Abeokuta)	100, 000	2.56

Total Billings for the period above   3, 905, 000

Table 4.2: An Advertiser's Media Billings (1990)
Note comparative disadvantage of OGTV in this

-165-

The pattern of advertising spend shows that Lagos-based stations were privileged; it also reveals the challenge stations faced in a competitive environment when advertisers had multiple opportunities to air their messages within the same market. Stations with the lowest share of this advertiser's budget were those serving rural audiences. However, NTA Channel 10 also had a low share, even though it was based in an urban area. It existed in the shadow of the more commercially motivated outfits, sister station NTA-2 Channel 5 and LTV. Clearly, unless OGTV was able to deliver the same quality of audiences as the Lagos channels, it did not have a chance. This explains why the management was keen on extending its reach beyond the state. OGTV was uniquely positioned to beam signals to Lagos, and much of the former Western region. Its signals were receivable across national borders in the Republic of Benin but at great costs. Had there been stricter regulation of the airwaves, the station should have been restricted to its assigned market in Ogun state. In this is evidence of the challenge of a commercial television station in a rural economy. In spite of all the odds, the station continues to thrive.

Like LTV, it had its travails during the military regimes. Military governors could rely on the federal establishment, rendering the stations owned by the state governments irrelevant. The extent of the disdain with which some military governors treated the station is captured in the threat to convert OGTV into a bakery! Little wonder that under the military there was an exodus of staff, especially to the private broadcasting organisations when these were established in the 1990s. They were lured away by the prospects of regular salaries and up-to-date facilities. The commercial orientation that they had imbibed was well aligned to the task ahead of them in the new organisations. Among those who jumped ship were top managers, commercial officers, producers and even newsreaders, those who had been the face of the station. A number of OGTV veterans joined the scores of independent producers, yet some remained, so the station kept faith with its rural audiences. It has since received a new lease of life.

When democratic governance was restored in 1999, the effective radiated power of Ogun Television's transmitters had been reduced from 40kw to 2kw, meaning it could reach only a fraction of its market. One of the first tasks of the civilian administration at the station was to purchase new transmitters, new generators and the Uninterrupted Power Supply facilities to back up the equipment. This was to avert breaks in transmission that had characterised the service. In addition, the premises were renovated and new production kit— cameras, non-linear editing facilities, compu-graphics packages—were bought. These changes led to the upgrading of the library; archival materials have been copied to DVD, which is now the mode for storage and distribution of programmes at the station. Staff have been trained in a range of relevant disciplines and at venues at home and abroad. In all these it is trying to keep up with globally acceptable standards.

The adoption of computers and global satellite mobile (GSM) phones in Nigeria has also affected the fortunes of the station. Though it had a relatively streamlined workforce to start with, management has reduced this even further with the elimination of secretarial and clerical posts. Instead, staff are equipped with laptops and an intranet service and functional telephones. The station is also on the internet; it lays claim to being the first Nigerian television station to maintain a presence on the worldwide web—www.gtvnigeria.com. This does not support reception of its programmes yet. Although the station has explored several opportunities for web-based delivery of its content, none had come to fruition in 2008. The use of GSM phones means that news on the station is more immediate—no more hours of delay, even with delivering international news. The newsroom is equipped with monitors dedicated to strategically selected global channels. As well as being broadcast in English, stories sourced from these are translated and used on news broadcast in the local languages —Yoruba and Oguu.

With the return of democracy in 1999, the station has been repositioned and re-christened Gateway Television GTV. It has a lot to show for this new image, in spite of the dreary situation with power supply that plagued television stations, their viewers and other businesses. It is proud of its award-winning coverage of the April 2008 elections, in which it was able to offer airtime to representatives of all registered political parties with a bonafide presence in the state to present their programmes to the people. That it could do this at no charge is remarkable for an aggressively commercial station as it used to be. This non-partisanship in a state-government station requires some explanation. The following is the insight offered by the assistant general manager in 2008.

> There is a difference between government and the governor. The governor is the one seeking re-election, we have to report government activities from Ogun state. We have to redirect our news angle and adopt a house style which will not make us liable to any laid-down laws and will not make us subservient to any party dictactes.
> (Demola Odusanya, AGM, Production Services interviewed in July 2008)

From this is it apparent that though the station is still commercially driven, there is also a measure of professionalism and maturity. There is limited financial dependence on government. For services rendered, it receives from the government a monthly flat fee and a tanker load of diesel. It also continues to nurture its streams of funding. There are marketing offices in every zone of the federation. The renovation of the premises and the construction of new halls affords the scope for a steady income from the hire of corporate events and private parties. A substantial amount accrues from these activities. Yet a good chunk of its revenue still comes from religious programmes, which it has tagged the Faith Brand. Religious organisations abound with messages to propagate; they, unlike government, will pay commercial rates for the airtime. Heavy reliance on this brand contravenes regulations on religious broadcasts, yet it is a lucrative aspect of television business. It may suffice for now, but for how long?

Gateway Television is already preparing for Nigeria's switchover to digital transmission, proposed for 2012. Whether the radical strategies adopted in marketing, programming and news reporting will suffice in the coming era will be the real test of how viable these are. In the meantime they have been proven, given the right political climate.

Plateau State Radio Television (PRTV)

Plateau Television can be said to have two incarnations. The first was when it began operations during the early phase of television in Nigeria; a project that came to fruition as Benue-Plateau Television

(BPTV), the first colour station in Nigeria. That station had a long gestation period. Its vision is attributed to the first governor of Benue-Plateau state, the late police commissioner JD Gomwalk (Mannok et al, 2005: 48). It is on record that after his achievement of establishing the state-owned newspaper The Nigeria Standard, television was the next challenge, since he had inherited a relay radio station that was adopted to serve the state. The calibre of people involved in the planning of the television station shows the high regard for standards from the start. Like Lagos Television, this venture was purely an indigenous effort, in the sense that there was no foreign partnership to give an initial boost. The planning team consisted of the chief engineer from Kaduna Television (now a bishop of the Church of Nigeria, Anglican Communion), George Bako, and two academically minded practitioners from Ahmadu Bello University, Madu Mailafiya and the late Egyptian-born Professor Girgis Salama, the father of broadcasting in the Middle Belt of Nigeria. Each of these men went on to make his mark in television broadcasting, just as the station that they helped to establish made its mark within the NTA.

Salama became the first general manager of BPTV (later NTA Jos). Having designed and built the station, he also had to organise and manage its operations (Salama, 1978: 21). BPTV began test transmission in 1974 and was formally launched in February 1975. By 1977 it had come under the unitary authority of the federal government television establishment. NTA Jos remains a station with an illustrious record for the technical feats it attempted: colour transmission, puppetry and animation. Little wonder that it was selected as the site for the NTA's Television College. But the nation's gain was the state's loss, proud as it was of how far its initial investment had gone. By 1980, following the return to civil rule, the desire for a state-owned station was revived.

The bill establishing PRTV was passed by the state House of Assembly. Technically, the new station had to serve only a fraction of the old Benue-Plateau state, which has since been carved up into more states in an effort to bring government closer to the grassroots. PRTV still had to contend with the configuration of towns and the topography of the state, from the highlands—Jos with its hills and towns like Pankshin on the fringe of mountainous ranges—to the wooded area of Lafia and Nassarawa in the south. The policy to restrict state-owned television stations to UHF transmission made this second attempt more challenging than the first.

The initial plan was to have a VHF station that would be completed in one phase at the cost of 12 million naira. As discussed earlier, adopting the proposals for UHF transmission meant upward revisions of the budget. This delayed the execution of the project, as the initial design had to be revised. The acquisition of the permanent site and completion of the buildings also led to delays.

In the end PRTV began operating on two UHF channels: Channel 29 for Jos and its environs; Channel 53 for the Shendam area of Plateau state. Its coverage area includes parts of Kaduna, Bauchi and Taraba states. This configuration reflects the boundaries of the old Benue-Plateau area.

The project was placed in the tested hands of Girgis Salama. He had 'midwifed' the first station but had returned to academia, from where he was recalled to become managing director of Plateau Television (PTV). As in Ogun state, television was independent of the state radio service then. PTV's aspiration for excellence was evident in the investment in training. From the initial tranche of money available to it, the station acquired state-of-the-art technology. It is on record that 62 of its staff were sent abroad (UK, USA, Canada and France) to a range of institutions to acquire expertise in production techniques and skills in using the equipment. Finally PTV was commissioned in 1982, backed by State Law No 1 of 1982. The launch was part of the events marking the national convention of the Nigerian People's Party (NPP) in Jos. PTV began its operation with the coverage of this spectacle. This coincidence illustrates the symbiosis between television and politics.

That its mission and fortunes were tied to politics meant the management of television was as unsettled as the political storms in the background. Following the initial four-year tenure when Salama managed the station, the station went through a period of upheaval. The changes reflected the shifts in political parties in government. The ruling party in Plateau state was the NPP. That NPP was one of the opposition parties, put the state on a collision course with the ruling party at the federal level - National Party of Nigeria (NPN). NTA Jos, which the people saw as their own establishment, was now under the NPN federal government, controlled so that it could no longer be trusted to serve the interest of the state. Like other state stations, PRTV adopted a parochial outlook. Dissent in any form was not entertained.

By 1984 the military was back in power and the station was placed under the charge of sole administrators. In two years, it passed through the hands of four of these. The military government brought the state television and radio services together as one corporation in 1985. This merger had an impact on staffing and funding. Anxieties accompanying the realignment had to be resolved. Over the years, the pendulum of concerns swung from access to updated technical facilities, to recruitment of proficient staff. Each station manager had to contend with these while also dealing with the nuances of political leaders and inadequacy of funds. The 1980s and 1990s were the decades of the International Monetary Fund-imposed Structural Adjustment Programme, and money was in short supply. It was also the period when plans for the service needed to be consolidated, with regard to capital projects such as works on the building for the joint radio/television operations. Management also had to consider how to establish an institution culture that it could be proud of. By the end of the decade some general managers had to contend with situations in which contractors could not be paid, yet the station still had the support of the Plateau people. With all that had been invested in the station, one begins to appreciate the apparent parochialism, a trend that continues.

NTA Jos, which is PRTV's direct competitor, suffers the consequences of the privileged treatment accorded to PRTV by the people of Plateau. When it comes to patronising the stations, the experience of NTA Jos is that Plateau people would rather support the state-owned station PRTV than any NTA initiatives. Some even point to the fact that the NTA struggled to get new community stations going in the state when the idea was mooted many years later. These observers attribute this to a lack of support from the state government and the communities (especially the business community). This is contrary to experiences in other areas, as attested to by the Zaria community station and the patronage of NTA Maiduguri, or NTA Makurdi. The fact that the same market attracted and still sustains private television stations when these began operations lends credence to the logic in this anecdotal evidence. If nothing else, it reveals another dimension to the politicisation of the media.

Other direct showdowns lend further credence to the story. At a point during the Third Republic (1998-2002) the Plateau state government antagonised NTA Jos. At the root of this were specific developments within local politics; a certain Plateau governor had crossed the carpet to one of the opposition parties, though he had been elected into office on the ticket of the ruling party. The hostilities that ensued permeated the social fabric and determined the lines that PRTV had to toe, especially in news reporting. Claims and counterclaims were played on PRTV and NTA Jos. This illustrates another reason why state governments insist on having their own stations. Television was a voice for the people, but it was also a theatre for political confrontation. Though it is frustrating, there is an understanding that this is the way to play the game. If the NTA succumbs to certain pressures, the state government-owned stations would also be "their master's voice". NTA stations can resist state governments (in a civilian regime), but state stations cannot. These conflicts seep through to programmes. Though it is disappointing that professionalism is mortgaged in these battles, some good comes out of the dual system of information. Discerning audiences are able to read through the alternative perspectives presented, especially within newscasts.

Television broadcasting has been subject to the insecurities of military leaders in Nigeria. It also became a hostage of the civil unrest that emerged with the awakening of religious and ethnic rivalries. Sadly, Plateau state, which had hitherto been remarkably peaceful, became a hotbed of political activism. Managing a television station in this state has proved to be a precarious business. Governors needed station administrators whom they could trust; individuals whose loyalty they could count on, even if they attempted to be professional. Because the appointment to the post was at the behest of the state governor, the chief was often replaced. This quick turnover did not help stability, because managers were not in post long enough to make profound impact on operations. Managers who did not do the bidding of their political patrons were starved of funds or simply replaced. So managers were reduced to tackling more basic but no less crucial aspects of their duties – resolving staffing structures, addressing community relations and issues of accommodation and the perennial challenge of adequate technology. PRTV was not immune from the neglect that state government stations experienced during the military regimes of the 1990s (see Mannok et al, 2005: 57-89). For the sake of survival, managers learnt to be tactful in walking the tightrope. There is an understanding of the need to find a middle ground between professional ideals and the demands of the political elite, as the following remark from a general manager of the station shows:

> We can't divorce politics from broadcast management. Just look at the reasons for broadcasting. Managers should be avowed professionals, they should hold their office in trust for the public. You should try to maintain a balance, giving government its dues, and the people what they need.
>
> To do this, you need to understand the challenges, policies and action plans of the government. These need to be propagated to the people. You do more to mobilise the people towards a common cause.
>
> (Yiljap Abraham, general manager, PRTV Interview conducted July 2008)

When the common cause is determined by the development and informational needs of the state, it is a win-win situation: for government, which desires to be popular; and for the people of the state, who require social change. The service is all the better if the effort is regarded as entertaining, informative and educative. Abraham cites two examples to illustrate this. Both happened in May 2007, when Governor David Jang was in office. Jos was very dirty, so dirty that it was said that when people from neighbouring Bauchi state passed by, they had to hold their noses. The government declared war on filth. PRTV took hold of that and ran with it. It was reflected in shows, news stories, editorials, commentaries and various programmes, even jingles.

> It was a campaign against the habits that build up the filth, it was called 'Doti na doti' [Filth is Filth and there is no excuse for it]. We showed how public utilities were being dwarfed by the heaps of dirt. We asked questions like: 'Is this a bridge or a rubbish tip?' We showed the Caterpillar [earth-moving vehicles] shifting the mounds

of dirt and gradually releasing the space [underneath the bridge that had been blocked by illegal dumping of refuse]. We followed up with images and slogans celebrating the change, saying, 'Things are getting better'.

Some of the programmes dramatised how clean the environment could be. There were reports of [how the] government employ[ed] widows [to provide them some wages and] to sweep the streets, and the results of such efforts. There was a jingle which showed a young man littering the portion of the street that had just been swept by the government-employed street-sweepers. The jingle reprimanded him and all like him with the catchy rhyme 'You throw, I throw, someone else throws, won't there be a heap of rubbish here?' We must give the people a common cause.
(Yiljap Abraham, general manager, PRTV Interview conducted July 2008)

The above is an example of television in its routine duties serving to mobilise the people. The second example is based on coverage of local and national elections. In recognition of the sensitive nature of the assignment, a workshop was organised by the National Broadcasting Commission. This was for everyone involved in the exercise, from the drivers to the directors. It included media professionals along with police, state electoral officials and from the Independent National Electoral Commission, representatives of civil society and the National Orientation Agency. A communiqué was developed that included a list of the needs.

The state was divided into zones and live coverage from the polling booths was organised. This dispelled fears that the elections were being rigged. PRTV was able to show that there were lapses but that these were not sectional. For instance, the shortage of polling materials experienced was not restricted to any particular sections of the state.

Based on PRTV reports and other information received, the governor cancelled the elections. The governor congratulated the station for a job well done; he had been monitoring the election coverage and was able to make a popular decision. That made me realise that we had earned his respect. The governor wants a station with integrity, a station that both the government and the people can trust.

The actualisation of this depends on the individuals in government and the professionals in the broadcast station. Broadcast managers must be politically sensitive but they must also be professional and strive for excellence.
(Yiljap Abraham, general manager, PRTV Interview conducted July 2008)

This is evidence of a fresh wind blowing through politics and the media. There is a new dimension to management. Station managers have added to their duties the task of managing their owners rather than simply doing their bidding.

If you are a professional, you will advise that a glut of information leads to lack of credibility. It will lower the rating of the government in the eyes of the people. When you [governors] talk too much, you arm the opposition and they will put more pressure on you. The ruling party should not be afraid of the voice of opposition. The opposition should be regarded as a resource for ideas. Be creative in responding to them, pick from their ideas. The audience will find you more credible when you do that.
(Yiljap Abraham, general manager, PRTV Interview conducted July 2008)

This sentiment was universally expressed in all the stations involved in the 2008 observational study. With this, it would appear that the issue of political control is being resolved. This leaves the other challenge—funding. Withholding vital revenue streams is one way governments restrict stations. Stations are more likely to find themselves at the mercy of adversarial government forces that control the purse strings. Stations have been known to be starved of funds to bring principled officers to their knees. In the process, quality of service has been compromised and audience satisfaction jeopardised. This should not happen if those in authority are reasonable, if they put the public interest above their political careers or personal concerns.

Television today must be aware of its need to add value to content. There is no point in merely giving information or instruction. A number of people have access to raw information from the internet

and there is no point simply replicating that. For television to remain relevant in these times, staff should be retrained to raise their awareness.
(Yiljap Abraham, general manager, PRTV Interview conducted July 2008)

Considering the high premium placed on staff training at the inception of this station, it is surprising that training was a constant concern; but an exodus of staff and changing technology make retraining essential. Over the years, there has been a need for technical changes, and stations have failed to keep up with development in the industry. As well as enhanced audience exposure, improved media literacy is further justification for TV staff to be challenged to raise their game. With greater access to more stations through the direct-to-home satellite channels, local stations have to compete with sleek productions from home and abroad. Stations are aware of this and station managers monitor practices on such stations, along with the output of their own staff. The conclusion is that the stations are subjected to globally determined standards of performance. This bears down on other areas of operation and is linked to the issue of training. It appears that the better-trained individuals are, the higher their expectations at work. In any case, it is only reasonable to expect that those who serve as advocates for a better society would expect reasonable conditions of service. They may leave the organisation in the absence of this. This is one explanation for the exodus of staff experienced in this station. This station's management is aware of this.

We need to improve the working environment. Trained staff have not been well cared-for, causing them to migrate to better-paying jobs. These factors have led to lack of continuity in the industry.
(Yiljap Abraham, general manager, PRTV Interview conducted July 2008)

In the final analysis, state-owned stations such as PRTV still rely on consistency in the political will of the executive, if their objectives are to be realised. With the symbiosis between stations and the state governments, the political will to sustain them is essential but policy-makers at the state level have not always been consistent in carrying a vision through. When faced with other capital-intensive priorities, politicians relegate to the background the demands of the stations that they set up. Successive administrations come with new agendas. Sometimes failure to act was caused by political leaders not understanding the nature and cost of broadcast technology or the broadcast business. Their ignorance was compounded by the culture of suspicion; in which they regarded bills as dubious and employees as thieves. This slowed the process of procurement of essential devices or spare parts. With the compromises that are made to keep the stations afloat, staff welfare often takes the back seat. Frustrated staff then leave for greener pastures, compounding the challenge in the stations. This shows the inadequacy of the heavy reliance on political patronage and government support.

Ultimately, real independence from political manipulation is financial independence. For this reason, PRTV (like other stations) tries to develop other sources of revenue. Beside its Commercial department, it has a promotions unit that is responsible for mounting events that will generate revenue from the sale of airtime and fees. These include music shows, fairs, exhibitions and the like. The station also offers commercialised services in technical aspects of production such as scripting, choreography and location scouting. This way it offers the best it has in order to get what it needs.

The organisation has a quality control unit that monitors its output, keeping it attractive. Even though Jos is not always regarded as a primary market, it is strategic in the national distribution network. With the goodwill it enjoys from its customers, viewers and listeners from its sister station, PRTV has been able to generate much-needed funds. In 2000 it recorded a rise in revenue of five to six million naira per month (Mannok et al, 2005: 154). To achieve this, the station had to change what was described as the administrative approach to the Commercial department and revenue-generation. A commission system was introduced, staff were given targets and there were incentives for sales and performance. As at OGTV, every member of staff, regardless of his or her unit, was a potential salesperson. Debtors were pursued more aggressively and the station pulled the plug on areas of wasteful spending. Sadly, it became a hostage of its own success as its monthly subvention from government was drastically reduced. It had demonstrated its viability.

In the meantime, its expenses had continued to rise, especially with added cost of power generation. The station also had to maintain a clear presence in Lagos and Abuja to attract the big spenders. It continued to play its part in industry initiatives, in the Broadcasting Organisation of Nigeria and National Broadcasting Commission's conference of African broadcasters, Africast. More importantly, it had to embark on a Turn Around Project. In 2004, two more television transmitters to boost the reach of the station were installed. Such improvements have become essential in the new competitive era as Jos has attracted independent television stations. But the station trudges on.

It has clearly articulated a forward-looking mission. PRTV seeks

> to be a formidable, innovative and creative broadcast outfit, through well-motivated and trained professionals; projecting peaceful co-existence and the scenic beauty of our land; effectively using our tripartite power to provide quality broadcast service, for the upliftment of the socio-economic development of Plateau State; bequeathing a legacy of agenda-setting and purposeful leadership for future generations and posterity.
> (Mannok et al, 2005: 151)

This mission statement reflects the challenges that the station has had to face, as much as its aspirations. What it does not capture is the serious pursuit of peaceful co-existence. Because they have seen their state weathering many storms, managers at PRTV are mindful of their role in keeping the peace. In spite of their best intentions, one threat remains, and it is evident in the programme schedule. Though the bid to bring the government closer to the grassroots is implied in the mission statement, it is ironic that the commercial imperative dictates that airtime is given over to more global and national agendas than to local ones. If capitulating to these pressures cannot be helped, hopefully the local news reports will maintain the local outlook and help fulfil the mission. This is a trend to watch.

Fig 4.14: A Glimpse of Plateau Terrain

Fg 4.15: PRTV cameraman in action on location

Fig 4.16: Reviewing camera shots on location

Fig 4.17: Control Room at PRTV

Fig 4.18: Cultural Performers from Plateau State

Fig 4.19: Local children in the shadows of PRTV mast
Images courtesy of PRTV Jos Management

Community Television Kano

The Kano state-owned television station is a community channel with a difference. Considering that Kano and Lagos have remained the most populous municipalities in the nation, having a community television initiative poses special challenges here. Community Television Kano (CTV) has a scope of coverage that transcends the vast locality of the ancient city of Kano. Its constituency extends to neighbouring states. With such a large population base to serve, it is unable to involve the community in the planning and production of programmes as intensely as suggested by development theories. Yet the station is deemed to be the people's television. According to its founding fathers, it is called Community TV because it is for the masses. Its programmes are designed not for the elites but for the people at the grassroots. It specialises in low-cost productions. Dramas rely on street performers who then become household names. Also, because of the low level of literacy in English, most of the programmes are in Hausa.

It is fully funded and owned by the state government. One could regard the taxpayers as owners of the station, and the station's name seems to justify this view but there is another perspective. The government in this context is seen as the individuals in office; those who sign the cheques to run the station. Even in 2008, the government still provided capital and running costs; funds for salaries, equipment and the acquisition of a digital transmitter, for example. The commercialisation wind that has hit the rest of the industry appears to have missed CTV. Clearly this community station is really a government mouthpiece, as its history confirms.

As in Plateau state, there were two attempts to establish a station for the people of Kano. The first was before the establishment of the NTA. By 1977, that station which was established to redress the limited reach of RTK had come under the umbrella of the NTA. That was under the military administration and there was no rivalry between the federal and state governments at the time. Changes witnessed at the dawn of Nigeria's second attempt at civil rule marked the shift. In the Second Republic (1979-83) states that had not supported the National Party of Nigeria (NPN) at the polls, those whose state governors belonged to other parties, fell out of favour with the federal authorities. This extended to the federal government-controlled organs such as the NTA.

Kano state was under the People's Redemption Party (PRP). The executive governor, Abubakar Rimi, considered the publicity he was getting for his government's initiatives on the federal television channel inadequate. Publicity is regarded as the lifeline of any government. Governments need to speak to their people, and to sell programmes of action. Without the support of people, social and political programmes may not succeed, and governments may not get the required votes to return to office, so access to information channels is crucial. The unwillingness of the federal station to help this state government communicate with its electorate compelled the governor to establish a new station, thus CTV was born.

Its operations were modest to begin with. The equipment was located inside Government House at the time. The target audience was the lowly illiterate masses, the primary constituency of the radical northern leader Mallam Aminu Kano, leader of the PRP; a reincarnation of the Northern Elements Progressive Union. Mallam Aminu had a tendency towards equity and social justice and his party promoted populist ideals. Although he was never able to secure enough support to win a national election, the PRP maintained a stronghold in Kano.

In the beginning there was a lot of entertainment (foreign musicals and Indian films) on CTV. Such programmes pulled in the crowds at whom the political messages were targeted. Perhaps the popularity of the programmes endeared the station and the government to the voters, the people who mattered. In this way it also began to carve its reputation as a community station, gradually getting to the point where 80% of programmes were locally generated. This was easily achieved, since the focus was on news and current affairs, children's programmes, religious programmes and local drama.

Changing scenes on the political canvas have not affected the orientation of the station as a mouthpiece of government. Irrespective of who is in office, the station still regards its primary role as that of the champion of government programmes. Staff have no qualms about being "the master's voice"; the master being whoever is governor.

> Of course it is [a mouthpiece of government]. Any reasonable person who comes on board will try to use it to sell [his programmes and ideas for governance].
>
> (Manager, News and Current Affairs, CTV Interviewed July 2008)

This is not always as sinister as it sounds. For example, CTV was used to promote the acceptance of the polio eradication programme. There were global ramifications to this local issue. The global initiative to eradicate polio was being undermined by rumours in parts of Nigeria, including Kano. The rumours attacked the motive of the eradication programme, with the result that fewer people than expected received the vaccine. A concerted effort was required to resolve this risky health problem. The station produced drama sketches and jingles to persuade people to accept the intervention in good faith and allow their children to be vaccinated. CTV also helps to alert audiences to other government policies and programmes. It works directly with state departments such as the Department of Children, Youth and Adolescents.

Nevertheless, Kano area still reported the highest number of cases of polio in December 2008. According to *Count Down*, a World Health Organisation newsletter, Kano had 272 polio cases. Zamfara was a distant second, recording 85 cases, and Katsina was next with 80 cases reported. This is a sharp contrast to the situation in Ogun (three cases); Enugu, Benue and Osun states (two each) and Adamawa, Ondo and Lagos states (one each). This gives an indication of the disparity in the challenges that television needs to address across the nation (or the achievements that previous efforts reflect). It is just one example. Clearly the priorities of the station must invariably change as social needs change. This would make for interesting television, if the focus did not appear to be constantly on people in government.

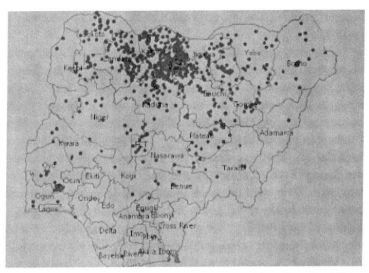

Fig 4.20: Map from Count Down *WHO Newsletter No. 6 December 2008
on Polio Eradication in Nigeria. Dots indicate the incidence of Polio
Courtesy WHO Nigeria*

The present civilian administration has given the station a free hand,
which encourages the station to conduct a dialogue between the
government and the governed. This way, the government has a better
chance of knowing what the people expect from it. However, this
was not the situation under military rule. The manager of News and
Current Affairs recalls that during the military era, news operations
felt more restricted. Not all stories that professionals considered
newsworthy could be reported. Messages were sent to the news
desk about stories which the military leaders did not want. Stories
were spiked on receipt of instructions like "We don't want that story!"
Or "Kill that story!" from military leaders or their representatives. It
was commonplace for such orders to come by phone. One view
suggests that the station experienced greater censorship under the
soldiers than it did under civilian administrations. This contrasts with
experiences reported by television broadcasters in other locations, so
it may be a function of the personalities involved. It is not clear if
the leaders in Kano were particularly touchy, or station managers
exceptionally cautious, but Kano is the home state of the infamous

General Sanni Abacha, Nigeria's military dictator, who held sway in the 1990s.

News managers justified their compliance with such requests as acts of social responsibility. A senior news person argued that "having peace was paramount. To avert the breach of the peace, it was better to look away." Journalists exercised a measure of self-censorship based on their understanding of their role under a military dictatorship in a developing country. This imposed tight limits on their duties. Perhaps due to the experience of the civil war, broadcasters buckled under pressure and made the pursuit of peace their goal. They were prepared to sweep certain things under the carpet to maintain social order, rationalising that this was part of their social responsibility. Yet secretly these pro-establishment public servants were not completely resigned to fate. They were tactful in their involvement in the struggle for democracy. They had their own means of rocking the boat, though not in a brazen way as the press did. While newspapers had cartoons, television used wit, sarcasm and craftiness.

> We get somebody who is highly respectable, an elder statesman, [possibly someone] who has participated in the struggle for the independence, we ask them about something [sensitive issues which need addressing]. [The interviewees are] people who are known to the audiences, and who are confident that they cannot be touched by the military. We approach those types of people and then we give them up-sound . . ." [sic – meaning use a sound clip from an interviewee]
> (Interview conducted in July 2008 with a news manager – identity withheld)

By asking provocative questions that will get this special class of people to express challenging views, the station acts as a ventriloquist, hiding behind "the sacred cows". When the sound-bite from the respected elders is heard, the station, in its timid way, has expressed its views.

Even as the mouthpiece of the government, CTV has adequate experience of being the underdog engaged in a struggle with the dominant power. First as the mouthpiece of the PRP-led state government pitted against the NPN-controlled federal might; then as a socially responsible broadcaster under the military dictatorship. Civilian rule as presently constituted is deemed to be the best for broadcasting. In the estimation of the manager, News and Current Affairs, "The present administration has given *open licence* to broadcasters." In any case, relations between broadcasters and the state (meaning the government and various sectors of society) have now been codified. The National Broadcasting Commission has published guidelines for broadcast operations which set out the parameters for judging an ethical political broadcast service. However, a new challenge to television broadcasting has emerged, and it is most evident in northern Nigeria. It can be described as a new form of censorship imposed by the dictates of a dominant religion.

Since Kano state has officially adopted Islamic law (sharia), CTV has to support the state's policies and help its audience to do the same. This is reflected in its programming. Islamic scholars (mullahs) are featured on the station to inform the people of their religious obligations. The station's contributions to society through its drama productions and light-entertainment programmes have to be moderated in the light of the prevalent religion. This pattern of voluntary regulation on religious grounds has also evolved as a critical concern of the station.

The CTV service caters for a range of audiences, including children aged between three and 10. They are regarded as a part of society who must be well integrated, so programmes are created for them. On the schedule are shows such as *Children's Court*. With such programmes, which feature the views of children on a range of topical issues, it would appear that the advocacy by UNICEF and the activities of non-governmental organisations clamouring for children's rights in Nigeria has been fruitful. The station appears to be helping children realise their right to participate and express themselves in their community.

The story is different when it comes to adolescents and young people. The station considers that this group is served through its coverage of the activities of the state Department of Youth. The presentation of the facilities and activities of this department is regarded by the station as being beneficial to society. Typically this includes programmes about adolescents learning trades, and street children being rehabilitated. This is regarded as part of the social responsibility of the station to report on the activities of government, part of its duties as a mouthpiece. In addition there are *Schools' Debates, Quiz Time* and other educational programmes. It may be due to the apparent formal approach to these programmes, or the finesse of entertainment programmes available on other channels, but the appeal of the effort in social responsibility is undermined. Whatever the explanation, it is clear that the station has an uphill task with this audience segment that would rather pursue trends on the global youth scene. Operatives are aware of the limited impact the station has on young people.

CTV offers programmes that should mould the character of the youth and encourage them to take pride in the local culture. However, in contemporary culture, there is competition from global sources through satellite channels and the internet. Young people have access to internet cafes, which abound in the city, and they go to viewing centres and cinemas that give access to alternative cultures, even those that television might censor. It is ironic that at its inception CTV had begun to cultivate the audience palate for some of these; now it prescribes another pill.

The station was very popular among youth audiences for the scores of foreign musicals that it aired when it began in the 1980s, when there was no manpower to make local programmes. For housewives and younger children, the Indian films featured on Sunday afternoons meant that some men had to forgo their lunches. These popular programmes won the hearts of the masses. That was when the station earned its claim of being the station of the people. With time (and the adoption of Sharia) moral guardians decried such fare as alien, corrupt and undesirable. Gradually the *foreign culture* was replaced by the type of fare sanctioned by mullahs. It can be argued that the phasing out of foreign programmes is due to a combination of factors: the discovery of cost-effective local productions; concern about the erosion of cultural norms and values; and the political will to use television to promote and preserve local cultures. The superior logic of using television to foreground respectable Islamic values (befitting a Sharia state) was partly due to the station's capability to accomplish this mission, but it also reflects the new political definition of the role of the medium in this section of the nation. The station in this instance has been adopted as a tool for moral guidance.

The station transmits for only part of the day—about seven hours. It is aware of the competition posed to its strictly regulated content, by new media technologies that are available round the clock. For 16,000 naira people can receive foreign programmes on Direct to Home Satellite Television (DSTV). A segment of the population visits the cinema or viewing centres. Nigerian-made movies (Nollywood, Kannywood) are readily available to watch at home. These make the task of satisfying the audience, particularly young people, more challenging.

There is strong anecdotal evidence showing that young people still watch MTV-style foreign music programmes (for example, on the South African-based Channel O via DSTV). They also follow British and European football leagues, which are aired on satellite television. It thus seems that the strict regulation of the content of CTV is able to do no more than to set boundaries so that audiences can at least know what is approved, even when they choose to transcend these. This internet generation will not be told what to do, if they can afford access to other media. Yet the resultant pattern of access to the media is likely to create further cultural gaps in the society.

If one were to go by the known programme preferences of young people, it would appear that being like other youths is more important than any distinctive ethnic or religious identity. Yet this audience segment should not be lost if the government's mission is to be accomplished. The youth are critical for the future, as expressed by one of the station managers.

> If we can tap them, bring them within our fold, we show them programmes to do with our norms and culture. Culture is very rich. It makes you dignified, makes you a proper human being. Anyone without it will be in the doldrums, having no legacy to leave behind. Television should protect, defend the norms and culture of its peoples.
> (Interviewed in July 2008)

Noble as this view is, it fails to recognise that audiences cannot be compelled to view what the station deems best. Programme preference is a matter of taste, and taste cannot be imposed. It has to be cultivated, even negotiated. Even if there were no foreign influences or alternative stations to choose from, the audience has a choice, to view or not to view. Therefore it must improve the quality of its drama programmes if it is to satisfy a broader range of audiences. The CTV experience suggests that messages do not thrive merely on their cultural value. Media messages are judged for stylistic and aesthetic appeal. With the continued domination by trans-national media corporations, such standards are set by forces far beyond the local vicinity. Audiences may not be as forgiving of perceived stylistic lapses in spite of the sentimental attachment to the station. The challenge is to find a means of situating messages within acceptable cultural contexts; making programmes that distance themselves from cultural taboos while still being attractive and desirable. Television requires money to hire professional artistes- whose fees the industry can no longer afford. It requires the production facilities and expertise to ensure a polished finish of the programmes. This is an expensive venture. It is much more than a business and must no longer be regarded as a town-crier. It is the struggle for identity, for cultural preservation in the face of pressure from attractive foreign alternatives. That television is a battleground for hearts and minds is more glaring in this context.

Notes

1. Having been ruled by the military for almost a decade, Nigerians, urged on by the international community, were anxious for democratic rule. The elections were conducted under the military regime of President Ibrahim B Babangida at a time when support for military rule was at its lowest ebb. The presidential election on 12th June 1993 was deemed by local and international observers to be the freest elections ever held in Nigeria. Chief MKO Abiola, a Yoruba man, secured broad-based and truly national political support, yet the result was annulled by Babangida, a northerner. Subsequent events led to the imposition of sanctions on Nigeria by the international community. June 12 became symbolic of the pro-democracy struggle and has even been set aside as a public holiday in certain states (See Falola & Heaton, 2004; Onuoha & Fadakinte, 2002).

Chapter 5

The Third Wave of Expansion Post Deregulation

After three decades, television in Nigeria had reached more homes as additional stations had been strategically located across the land. State governments, along with the federal government, had stations that they controlled. These stations had made bold strides, chronicling the social existence of the people and showing new ways of life while celebrating the old ways. As should be apparent from the case studies, by the end of the 1980s a reasonable proportion of television stations appeared to have been neglecting the interests of the audiences in their attention to the people in government and government activities. This was symptomatic of the neglect and discontent apparent in the land. Audiences, both the urban elite and the less enlightened desired a better quality of life than was locally available and some went out to get it. So great was this drift that government had to embark on a campaign to keep people from emigrating in droves. The message was that "the Andrews" (a term used to refer to émigrés) must not check out; Nigerians must be encouraged to invest in their country. Private television stations were ushered in during the advanced stages of this discontentment. This chapter will present some insight into the events that led up to this situation and examine operations of two examples of such private stations as case studies

Nigeria's petrodollar earnings had helped leaders to raise the hopes and esteem of the nation; sponsorship of events such as the 1977 Festival of Arts and Culture (FESTAC) earned it the designation "Giant of Africa". Nigerians had also become more exposed to Western (so-called modern) lifestyles and aspirations. Television had played a part in creating awareness of these. This nation was great not just because of the size of its population, but because it was also able to do great deeds, which people could see on television and in other media. However, Nigeria fell on hard times. Government was no longer able to bankroll many of its projects. It had gone cap in hand to find help—to the Organisation of the Islamic Conference (OIC), of which it became a member in 1986; and to the World Bank/International Monetary Fund (IMF), which prescribed a Structural Adjustment Programme (SAP) that it adopted in 1986. Such alliances shaped public-spending and social policies; this was the era of conditionalities. To the horror of many Nigerians, prices were hiked; the price of fuel, and by extension the cost of transportation, and almost everything else went up. The naira was devalued. Inflation took its toll, on individuals and organisations. Popular culture was rife with images of how to combat impoverishment. There were contradictions in the range of realities presented in the media. Going by the representations on television, living abroad in America and Britain seemed like options to be preferred; urban Nigeria seemed to offer a better quality of life than the rural areas. With such dichotomies, people from various stations of life were discontented and could cite evidence from the media to justify why. Nigerians were disillusioned by the contradictions evident in local, national and global life (Esan 1993).

By the 1990s, global trends and externally imposed economic policies (notably the SAP) had prompted government to encourage private-sector participation in education, health and the provision of roads and other infrastructure. This revision of government attitude to its control of public utilities was in part responsible for private participation in and commercialisation of the broadcasting industry, as noted by Atoyebi (2002: 15).

This was the situation in 1992 when the National Broadcasting Commission was established by Decree No 38 (amended by Decree No 55 of 1999). These legal instruments equipped the commission to regulate broadcasting; to consider applications for private broadcasting licences and recommend those to be approved. The final approval was still vested in the head of state, who at this time was the military president, Ibrahim Babangida. This was on the eve of the aborted Third Republic (the annulled 1993 elections -.see endnote in chapter 4) Political aspirations were high, and there were strong sentiments in favour of democratisation. A new wind was blowing across the land. The political will was supportive of deregulation of broadcasting.

Several reasons have been adduced for this support (Saidu: 2002: 27). The summary of these is that government came to review its role, accepting that it should be a custodian rather than owner of the airwaves. From this emerged the higher regard for the need to reinforce broadcasting's potential, to make it foster national integration and development, especially in rural areas. It was acknowledged that broadcasting is vital in promoting cultural pride, and that access to locally relevant programmes was one way to counter the allure and influence of foreign programmes. These arguments were based on the need for a free flow of a wider range of ideas, particularly from within the local context. One can infer from these arguments an acceptance of the limitations of the government-owned stations; thus it would appear that by acquiescing to the participation of private broadcasting organisations in the mediascape, even those in authority had recognised the need for a shift. This gives the impression that the pursuit of the democratic ideal was paramount. To regard these developments as isolated local trends would be naïve. Deregulation of broadcasting was part of a global trend. This was the era that marked an exponential growth of channels in nations around the world (Preston, 2004: 46). Likewise it was the period when global channels became firmly established, with brands like MTV and CNN being distributed via different platforms across the globe (McMurria, 2004: 38). Nigeria was not immune to this trend, as the national broadcaster was one of those that delivered their audiences to the global media giants in the early 1990s, when, for a brief period, CNN programming was broadcast on the National Television Authority (NTA). This was an advancement on previous practice, in which globally syndicated programmes were featured on air, as described in an earlier chapter. With these, Nigerian audiences were removed from their immediate locales, stretched across time and space in their mediated experiences, resulting in what has been described as a "disembedding of the social systems" (Moores, 1997: 238). This was cause for concern, yet old broadcast regulatory regimes with their national outlooks had become anachronistic, since nations could not withstand global torrents from trans-national media corporations.

The economic logic of reaching bigger markets that accompanied new technological capacities of the media could hardly be resisted. Some entrepreneurs were of the view that the only way to beat the trend was to join in the game. It would take people with an entrepreneurial mindset to deliver viable service.

Though Saidu presents deregulation as government's awareness and respect for citizens' right to participate in the use of the airwaves, the conditions for participation still meant that the medium was the preserve of the privileged. Application forms alone cost 10,000 naira, and the fees for a five-year terrestrial TV licence were 1.5 million naira in 1993. These had gone up to 50, 000 naira for the application and 15 million naira for the licence by 2002. Although the system was in place to make broadcasting more independent of government, it was clear that the private entrepreneurs who would break into the liberalised industry were those who had the means and the clout. No wonder there were divergent views about the merits of deregulation. While some people celebrated the promise of creativity and pluralism in broadcasting, and some enthused about the prospects of independent news production, some remained sceptical about the compromises to quality and professional standards. Worse still, some Nigerian scholars like Sobowale and Uyo (quoted in Sada, 2002: 99), recognised that independence from government influence was merely going to be replaced by another dominating force—business. This faceless force was just as potent and insistent as government officials —after all, money talks. It was not always devoid of political dimensions, either. This was clear right from the start.

There was no shortage of applications even in 1992, as is indicated in the notes of a national regulator.

> On 10th of June 1992, presidential approval was granted for 14 television and 13 cable re-transmission as the first private broadcasting stations in Nigeria.
> So eager were private entrepreneurs to venture into broadcasting that a radio station went on air for test transmission before being formally licensed in June 1994. The NBC appropriately ordered a shutdown, after which government reconciled and regularized its authorization by August 1994. . .
> (Atoyebi, 2002: 10)

The audacious move is evidence that this was indeed a new era. The power of capital was asserting itself in brave new ways. Prior to this, professionalism had to contend with technology, government and politicians, but new influences, particularly commercial interests, would be tested in the new era. These would further shape industry practices. Among other early players were Clapperboard Television, Channels Television Lagos, Minaj Television Obosi, DBN Television, Africa Independent Television (AIT), Murhi International Television (MiTV), Galaxy Television Ibadan and Desmims Independent Television (DITV) Kaduna. The period has been marked by three types of initiatives: private stations, community (grassroots) stations and international service. The following case studies are presented to document the experiences peculiar to Nigerian television at this stage.

Channels Television

Channels Television was one of the pioneering private television stations. Its licence was granted following the deregulation of broadcasting in 1993. It commenced operations in 1995 as a news station, two years after getting its licence. There was much scepticism. Could a news station succeed in Nigeria? Only a decade before, similar concerns had been raised about the viability of CNN in the USA. The Nigerian situation, in which the definition of news had been curtailed, compounded the doubt.

Tight control of broadcast news was the norm on government-owned stations. News on those stations could largely be defined as uncritical accounts of what leaders do. That was more so in the years under military leaders who would not tolerate any dissenting viewpoints. This meant that audiences were sceptical of news items which paraded personalities and subjects that they resented within their homes. Consequently, it was difficult to imagine how audiences would be drawn to any station that was going to offer more of these "unwelcome guests". Channels saw the opportunity in this situation. It was the only private station floated by professional news people; it was an opportunity to turn the tide and show an alternative model of the news. John Momoh is the founder and chief executive of the station. He explains his motivation in the following excerpt from an interview with him.

> I thought there was a need for us to give people a voice, to let their voice be heard by government, and at the other end, to let government be accountable to the people. By so doing [to make government accountable and] serve the watchdog role that we [the media] are supposed to. Again, [we need] to let people know what their duties and obligations [are] to the state.
> (John Momoh, interview conducted July 2008)

Among his peers, Momoh is regarded as a visionary. He has worked in broadcasting all his life, starting on radio (Ogun State Broadcasting Corporation — Ogun Radio; Federal Radio Corporation Nigeria; Voice of Nigeria and later on the NTA as senior reporter, news editor and news anchor. By the time he started Channels, his was already a household name from the second wave of television. Besides his training on the job, he has a university education and has continued to improve himself, studying across disciplines in Nigeria and abroad. He is thus able to appreciate trends in the industry.

In establishing the station in such an unusual niche, Channels had rejected the generalist orientation to broadcasting that was prevalent at the time. This was a strategic choice, because news occupies a higher place in the programming structure. This is consistent with the observation made by Esan (1993) that news takes up more time on Nigerian television stations than any other genre.

News gives the station identity. Network news was the melting pot for all Nigerians. It has a serious impact on people; news is still an important service to the audience, especially if it can be properly defined and packaged. The choice of this niche area enabled the station to match its strength, the expertise of the founder and the core team, with the gap in the market. However, at the initial stages many people in the industry remained unconvinced of this potential. Momoh recalls being *chased out* of boardrooms by advertisers who were dubious about the viability of the station. How large an audience could a news station ever hope to deliver? The venture appeared to be too risky. It could end up being a waste of ad-spend. However, the station did not go under. Within three months into the operations, it had managed to convince the cynics that news is an appreciated service. Channels Television has since shown that news can be interesting and relevant, more than a showcase of prominent government officials; that Nigerians can pay attention to more serious news and current affairs programmes, and advertisers can reach audiences through these vehicles. Channels Television has turned the tide around. Its flagship programme, *The News at Ten,* captures a lot of attention.

It is regarded as a reference point for those who need information about the nation, and people have discovered that news is everything. Many now pay attention to information, knowing that information is very important . . . it's very important that we receive accurate information for us to make informed decisions. News is everything, our lives, you know, everything we do ... It will determine whether you go out in the morning or not, whether you go to the island or to the mainland — in case there is a riot, you know where to go to. That is why I am surprised when people say they do not watch local television.

(Teresa Essien interview conducted July 2008)

Having been raised on television, and with their level of exposure, Momoh's generation of broadcasters were critical of their practice even when they were operating within the constraining framework of government-owned stations. This was the generation of broadcasters who took on the task of monitoring gaffes made in newspapers in their *Boo-Boo Corner* (a segment on a newscast). They could identify with more critical (ABC) audiences as anecdotal evidence suggests, yet their reach cut across the spectrum. The AMPS Media Planning Data for 2008 shows that even audience with a lower class profile (C2-E) keep faith with Channels Television. The low incidence of audiences in a higher bracket may be due to the loyalty to other (foreign) stations. It may be that the more enlightened audience segment of decision-makers and people who want high standards have other means of receiving these on global news channels such as the BBC, CNN, Sky News and Al Jazeera. . With the historically poor reputation of Nigerian television news, it is little wonder that certain audience segments snub local channels in favour of foreign news channels, relying on papers to get local news. Yet evidence of loyalty from audiences in lower brackets attests to the level of engagement that can be generated across the social spectrum. Success with this group was not without costs, though. The station has been criticised for not having enough entertainment programmes. Some observers complain that Channels could support the Nigerian film industry (Nollywood) more, by highlighting its productions on its nightly *Movie of the Day* slot instead of showing foreign films. The station's practice is, however, evidence of its discipline and commitment to its defined focus.

In terms of programming, Channels took a fresh tack on family viewing from the beginning, by courting children who encouraged adult members to sample the station. It endeared itself to children with contemporary cartoons; adults were baited with late-night films. The News and Current Affairs programmes were "hammocked" between these entertaining shows and other documentaries. This was no unique formula but at the time, when foreign programmes were scarce on other free-to-air stations, Channels was serving up an irresistible selection of imported films, earning a reputation for being high-brow because of the lifestyle associated with Western fare. Yet it still devotes 76 per cent of its airtime to local programmes. This is an achievement.

Channels has been able to sustain its high local content ratio through its strategic programming orientation and prudent management of resources. As a news station, it has a high synergy between the news packages and the programmes. It generates material for some of the news programmes from research and interviews conducted in preparation for its rolling news bulletins. For a one-and-a-half-minute news story, a station may require 30-minute rushes or interviews conducted for that particular item. This would ordinarily generate no more than a few seconds of sound-bite and a few minutes of footage to accompany a news story. Yet the cost of getting the interview, which in many cases includes the cost of travelling to other parts of the country, is maximised when the rest of the interview is used from a different angle and discussed in more depth in a different programme, such as *Politics Today*. This can be described as *creative recycling*.

From its inception, the station has challenged the traditional concepts of news in Nigeria, broadening its remit to include new thematic areas. By adopting a range of current affairs programmes, Channels joins audiences in lingering on the events and issues that they care about. In 1996 it introduced the first television news programme on health, called *Health News*, which helped to dispel certain myths and create an awareness of healthy lifestyles, arguably contributing to an improved quality of life. The station relies heavily on its live programmes, *World Today, Sports News* and *Business News*. Politics, sports and economics sectors, which are highly regarded, are the station's "bankers". The decision to cover these areas is tactical. The business and finance sector is very strong; the station understands that it has a duty to report the economy. The enduring popularity of sports, and the presence of all-year-round fixtures, guarantee a constant supply of interesting reportage. Finally, the global news programme *The World Today* is a window on the world. Apart from international politics and economic affairs, it offers light-hearted stories, such as the one about "a panda is marrying a chicken", mentioned by Teresa Essien. In addition to these, there are 11 news-based shows per week each with different thematic interest. The station is proof that the programme production process can be tailored to tap into the collective wisdom of the team, as has been the tradition on older stations. Production meetings help to ensure that the diverse interests represented in the audience are catered for. Ideas and plans for the programmes are considered at weekly staff meetings. Channels Television has thus developed a rapport with its audience and a reputation as the station to turn to, to verify the legion of stories and claims making the rounds on other stations or in the rumour mill.

The vision for Channels was spurred on by the passion for broadcasting and the need to improve what NTA had to offer. This suggests that the NTA structure was stifling as the founders had been staff of the NTA. The Momohs (John, his wife, Sola) and the team were privileged to be able to realise this dream. Although many people had had innovative ideas about the television industry, few had the privilege of assisting at the birth of the dream. It would not have been possible without the Babangida administration's deregulation of broadcasting. This was the right era for private entrepreneurs to engage in the broadcasting industry, with the right policy in place and more affordable technology. Some had grand plans from the start, but Channels took the slow and steady route.

It began as Channels Inc a consultancy firm and producer of documentaries. From this came Channels Television, which started humbly, working out of rented premises in a high-rise building, unlike its early competitors who began their operations in purpose-built studios. It was able to do this because of its streamlined workforce. The height of the building reduced the outlay on transmission masts. The station was more visually appealing because of improvements in technology available in the 1990s. These enabled it to achieve sharper, crisper pictures despite more compact operations.

About 14 years after it started, a phased expansion of the business is unfolding. The station has established operations in the Federal Capital Territory, Abuja, with plans under way for bureaus in 15 out of the 36 states, including the cities of Jos, Benin, Port Harcourt and Kano. The properties for the station's permanent site have been negotiated for, and Channels will eventually move to its own premises. Whether this will have the unwanted effect of embroiling management in infrastructural distractions, undermining the core business of television, remains to be seen.

It took more than empty determination to break away from the NTA model, which was to ensure that the news station was "acceptable". To its credit, Channels brought a fresh perspective to news, by paying particular attention to the craft of presentation. Regardless of the degree of formality in presentations, elocution was as important as appearance and journalistic values. The Channels attention to professional standards is evident in its programmes, from the more serious *Abuja Weekly, Business News* and *Health Matters* to shows with more relaxed styles such as *Art House, Metrofile, Sunrise Daily*, the daily news and current affairs breakfast show, and *Sunrise*, which is the interactive audience participation show on Saturday mornings. To complement this freshness, it adopted a policy of introducing new faces to the screen. Young graduates were hired and trained. The station takes pride in its track record, and its professionalism in its core business has been widely acknowledged. For example, the principle of balance was defined more broadly at Channels; it was the first station to commit to getting not just two sides, but all sides of a story. It is known for being a place to hear alternative viewpoints to those disseminated on government-controlled media, views that would not otherwise be allowed within the mainstream.

During the struggle for democracy, Channels' practice of taking on unexpected stories helped to give its news the vibrancy which the Nigerian press had, but which was lacking on television. In spite of the treacherous experiences that accompanied radical journalism under the military regimes, Channels fared reasonably well. With its professionalism and non-partisan orientation, it was able to weather the storm of the military dictatorship. Everybody had a say on Channels Television, and it earned a reputation for being honest, fair and open. For example, in a move that went against the grain of popular opinion, Daniel Kanu had his day on Channels Television, laying out his case for adopting the military dictator General Sani Abacha as a civilian president of Nigeria. Kanu was the president of the now-defunct Youth Earnestly Ask for Abacha. Those opposed to the move were also given ample airtime. The station's policy was to avoid censoring views simply because they were from the "wrong" side of the political spectrum. Censorship was effected only when views were obscene, libellous, illegal or distasteful. This disarmed potential critics. Channels thus acquired a reputation for guarding the interest of the people; a station that the audience could trust. In this it has demonstrated that professionalism is in itself a weapon of defence.

There are many risks attached to the job of a broadcaster. The general breakdown in security is a major societal problem. The inclement hours associated with broadcasting heighten the risks to practitioners. Yet they were more risks in the course of duty. In the days when military coups were frequent, TV continuity announcers and newscasters learnt to cope with the gun-toting soldiers, either as coup planners or as guards at the stations. These days broadcasters face new threats because of their defined missions. Channels' staff, vehicles and equipment are vulnerable to attacks. Sometimes they are set upon by the very people they are meant to serve. News teams have been assaulted; cameramen are most vulnerable as they struggle to protect their kit while being beaten up by thugs.

Chimezie Obi Nwaogu, a Channels reporter, was invited to cover a story on a local land dispute. She arrived at the location in the middle of nowhere with her cameraman, and the complainants were nowhere to be found. Instead they encountered the locals and thugs, who made it very clear that the television crew was not welcome. The reporter, because she was unencumbered, was able to mingle with the crowd, escaping to safety. The cameraman was not so lucky. He was beaten up; he was holding on to his camera while being punched. In the fray the camera got damaged.

Teresa Essien, on what would appear to be a benign health reporting beat, had a similar experience. On this occasion she was sharing the camera crew with another reporter, Rosemary Omose. Essien's assignment (a report on a doctors' strike) was to come first. It was meant to be a brief stopover at the hospital. The medical director had other ideas. He was not having any news crew on the premises, as he preferred to deny that the strike action was happening. He threatened to unleash the security dogs on them. Essien recalls that the dogs were allowed to get within inches of their feet, baring their fangs. They had been obstructed and harassed. They were taunted this way for several minutes before other doctors managed to convince the "big man" to let them go.

Deji Badmus won the Television General News – News Bulletin category of the CNN MultiChoice African Journalist award (2008) for his report on the Third Mainland Bridge in Lagos. In the course of investigating, he came close to being thrown off the bridge by hoodlums (known as the Area Boys) who felt threatened by the presence of a television camera. With their reputation for using illegal substances, there was good reason for the Area Boys to be paranoid and to avert scrutiny, even though on this occasion the report was not about them. The reporter was able to explain his mission and reassure them because he spoke the local language. But not before he had been rattled with the fear of being thrown over (along with a used tyre; with a bit of luck he could stay afloat in the lagoon till he was noticed and rescued). To appease his captors, Badmus had to part with a thousand naira before he was released. These are some of the hidden costs of production and causes of disruptions to the production schedules. In the Nigerian environment even supposedly inconsequential or mundane reports can be risky and eventful. There are no insurance policies to cover all eventualities in this job.

> Those threats can be very frightening and they can be very real because you are there and you are alone, there is no backup anywhere. It's

[only] your negotiating skills [standing between you and the danger].
In Channels there is a policy that at least one person in the crew must
speak the language of the area being covered.
(Teresa Essien interview conducted July 2008)

As discussed elsewhere in this book, Nigeria is linguistically diverse,
so such a policy may create operational difficulties. However, it is
informed by the vulnerability of individuals when they do not
understand the language in the area where they are covering an event.
It is important for reporters to be vigilant and aware. There was the
case for a reporter on assignment in the Katsina area, who charged
along in a quixotic manner in pursuit of his story. He was oblivious
of impending danger because he did not speak the local language.
But for the vigilance of the camera person, who was proficient in the
Hausa language, another news team would have fallen prey to a crowd
of agitated locals who sought to resist the intrusion of strangers –
the media. All these have to be factored in when television stations
plan their business. How, for example are assignments covered? As
Channels expands its operations, setting up bureaus around the nation,
these and others are definitely issues to be considered. Though it is
not unusual to hear of journalists being beaten up, especially on the
orders of those in power, this trend indicates new manifestations of
power in society. Besides politicians, there are others who take power
into their own hands. With increased awareness of the power of the
media, some attempt to prosecute their cases in the court of public
opinion, dragging the media into their broils (as in the coverage of
the land dispute). Television camera people are more vulnerable. The
television camera makes reporters conspicuous; they are less able to
make a quick getaway from a spot of bother. In the event of an
attack, the concern is as much for the camera as it is for the bodily
harm done to the reporters. Broken cameras undermine an already
modest stock. The loss of even one out of 13 cameras is too great
an inconvenience for any station. Because equipments have to be
repaired abroad, such incidents are extremely costly, depending on
the foreign exchange rate and there's no insurance cover to fall back
on.. Still the station has many awards to attest to its persistence and
professionalism: it has been voted TV Station of the Year four times
between 2000 and 2008.

As most of its programmes are generated in-house, Channels is able to place its mark of quality on the bulk of its fare. Over the years it has placed a high premium on the excellence of its staff, whether they are on or off air. Channels tries to replicate the strict standards of presentation that Radio Nigeria (Training School) instilled. It balances recruitment of tried and tested hands with grooming of promising newcomers. Similarly, it maintains strict professional standards in reporting. Channels staff are thus likely to be an asset to any station. For this reason they are subject to enticements from other proprietors, using higher pay offers. Staff have come to appreciate the value of the personal development that Channels offers, as well as the realistic, possibly modest but promptly and regularly paid salaries. Still, when there is intense pressure from outside, the temptation to succumb to offers from other broadcasting houses may be high. Through its consistency and experience, Channels has been able to withstand attempts by the competition to deplete its staff base. This suggests that, with time, practitioners, like the television station, will define and pursue their interests. The nation may expect higher levels of professionalism when interests and not stipends become the driving force in careers.

Prudent management is the forté of the parent company Channels Inc, which offers its expertise on a consultancy basis. It prides itself on being the most professionally run station. Though it has a high regard for professionalism, and manages to keep regulators at bay, it has had its brush with the authorities. In 2008 its operations were temporarily suspended by the national regulator, the NBC—evidence that running television is still a minefield.[1] In spite of this isolated incident, Channels is better known for its scores of local and international award-winning reports. In 2006 it was rated the most reliable and trusted source of information in Nigeria by BBC/Reuters. In this way it strives for excellence, not succumbing to the temptation of using the so-called Nigerian Factor as an excuse for mediocrity.

The station has also demonstrated its leadership potential by attempting to establish cooperation among Nigerian and other African television stations. Channels Television was the rallying point for *TV Africa*.[2] Other stations involved in the proposed continental arrangement could have benefited from this programme supply chain, and perhaps the commercial value of the African market would have been enhanced. Sadly, the arrangement failed. The most obvious reason for this failure was the dispute over the rights to the African Nations Cup football tournament in 2002, but simmering in the background was the clash with the national regulators, who objected to the unofficial alternative network that was emerging in Nigeria (Salihu, 2002:168). It is inconceivable that this was a cogent obstacle to the success of a continental initiative. Surely such issues can be negotiated and resolved? Perhaps the industry was simply not ready for such an initiative. With time Channels may begin to pursue its vision. Channels is currently on UHF terrestrial television, but it is also on the West Africa Bouquet of DSTV Cable Network and on the internet (www.channelstv.com), though its programmes are not available for viewing via this.

The management at Channels is trying to ensure that with planned expansion, standards are not allowed to slip. It is aware that there is no one-size-fits-all formula, so as it expands, a relevant workable strategy will be evolved for each area. What it seeks to ensure in each case is that its programmes are people-oriented and the station remains professional, with its focus on news and current affairs. Each of the stations will, however, retain a distinct flavour, reflecting the identity of the locality it serves. There are plans for the new stations to carry some of the programmes from Lagos occasionally. This appears to be another way of fostering an exchange of ideas within the nation, something the network facilitates. This may be a worthwhile experiment to contrast with the existing orientation to national integration.

If it gets a licence, Channels will consider going network but it has more ambitious goals which should be encouraged. Like some other Nigerian broadcasters, it plans to get on the B Sky-B platform but unlike these stations that have a general programmes format, Channels Television wants to be known as the African Voice for News. Besides these general programming stations, there are other Nigerian specialist channels on the Sky platform, operating in the movies and religious belt. Channels television is poised to blaze the trail in this area, to beam the news to the UK and Europe. This vision is more attainable in an era of deregulation. Yet the magnitude of the resources required for this task, and the challenge of reporting across Africa on a regular basis, are unprecedented. Channels' news operations to date are encouraging, and it may well deliver on this goal. However, regardless of the enhanced opportunity for a south-north dialogue offered on the Sky platform (UK), there is still a long way to go in the bid to reverse the information order simply because audiences are fragmented. With the array of channels they are able to choose from, the fact that there is an alternative digital platform (Virgin Media) or even because of the deterrent to subscription due to costs, signals on digital television abroad may well be in a ghetto. Chances of reaching audiences other than Nigerians in the diaspora are still quite slim. Perhaps the prospects of reaching wider audiences will be improved if international channels could be received on free-to-air digital platforms such as Freeview, but these are issues beyond the control of local broadcasters. Nations reserve the right to regulate their airwaves so current arrangements must suffice for now. In the meantime, as Channels Television has demonstrated with its brand of private television, the first step required in fostering greater confidence in the sector is commitment to professionalism and a defined vision.

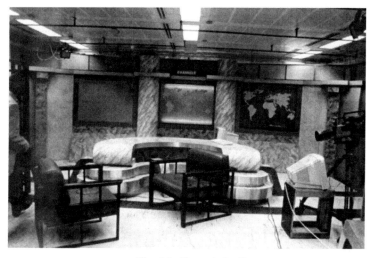

Figs 5.2: Channels Studio

Figs 5.3: Tracking quality of output

TV Stations Watched Most Often	Total	Male	Female
Total	76,618,125	37,963,806	38,654,319
None	18,165,068	8,738,383	9,426,685
LTV8	820,376	328,115 40%	492,260 60%
GTV CH 25 Abeokuta	804,015	427,196	376,819
PRTV Jos	518,973	272,248	246,725
CTV 67 Kano	642,679	338,995	303,684
Channels TV Lagos	492,028	301,938	190,090
AIT Lagos	1,271,122	695,793	575,329
NTA CH 4, 5 & 7 Ibadan	696,967	354,597	342,370
NTA Enugu	1,253,146	608,164	644,982
NTA Kaduna	199,325	112,226	87,098
NTA Kano	1,243,379	759,139	484,240
NTA Jos	250,979	140,378	110,601
NTA Zaria	652,265	332,371	319,894
NTA Abuja	172,449	63,783	108,667
NTA 2 CH 5	61,849	31,935	29,914
NTA CH 10	50,899	22,436	28,463

Table 5.1: Most Viewed Stations by Gender
Performance of Selected Stations - Courtesy AMPS Media Planning Data 2008

TV Stations Watched Most Often By Social Class	Total (Auto-base)	A	B	C1	C2	D	E
LTV8	820,376	7,479	40,812	34,577	367,978	348,546	20,984
GTV CH 25 Abeokuta	804,015	14,957	18,341	50,275	347,631	304,974	67,837
PRTV Jos	518,973	0	8,508	25,523	110,601	353,071	21,269
CTV 67 Kano	642,679	5,043	14,636	5,043	224,967	202,833	190,157
Channels TV Lagos	492,028	0	7,479	80,261	285,860	103,471	14,957
AIT Lagos	1,271,122	7,479	54,733	149,139	716,549	277,144	66,077
NTA CH 4, 5 & 7 Ibadan	696,967	0	30,569	158,957	232,322	256,777	18,341
NTA Enugu	1,253,146	4,970	98,064	94,905	565,429	411,150	78,629
NTA Kaduna	199,325	0	18,212	18,212	75,211	78,583	9,106
NTA Kano	1,243,379	15,129	29,273	69,124	510,439	315,735	303,679
NTA Jos	250,979	0	12,762	21,269	89,331	123,362	4,254
NTA Zaria	652,265	0	18,212	13,659	245,864	333,553	40,977

Table 5.2:: Most Viewed Stations by Socio Economic Classification Performance of Selected Stations - Courtesy AMPS Media Planning Data 2008

African Independent Television

DAAR Communications, the owner of Raypower radio stations, is an organisation with a dynamic goal. Raypower is the first private radio station in Nigeria. It aims to "establish, develop, maintain and sustain a global business institution that will be a market leader and offer an enduring platform for professionals to actualise their career objectives as consistent with corporate aspirations". African Independent Television (AIT) comes from this ambitious stock. From its name and the goal of the parent company, one gets a sense of AIT's mission. The vision set out by DAAR Communications is that this organisation will "promote the rapid integration of the African continent with the rest of the world; encourage development in Africa and attract development [opportunities] to Africa, project the hope and common aspirations of the black race and advance the positive goals of humanity. To share everything positive about the African experience . . ." (DAAR Communications, 2008)

These aspirations, like those of the Nigerian Television Authority, reflect the nation's foreign policy, which has Africa as its centrepiece. This demonstrates the commitment to issues of national development, racial and global politics and the local and world information order, even within the private sector. AIT has been unequivocal in its desire to bridge the gap in the world information order. Its mission is to "Share the African Experience", as expressed in its payoff line. Its proprietor was impressed by CNN's Ted Turner and Rupert Murdoch. Nevertheless, his initial interest in broadcasting appears to have stemmed from his experience in business and politics. One would expect that the organisation would be run efficiently, if not with a profit motive, if it is regarded as yet another capital-intensive business venture. But does the organisation run a service distinctive enough from the national broadcasters? In this section, AIT's model of television is considered.

AIT began in 1996 at its Ilapo village premises in Alagbado on the outskirts of the Lagos metropolis. By adding the cost of providing basic infrastructure (buildings, roads and electricity) to the technical facilities for the core business (transmitters, camera chains, vehicles and expertise) DAAR Communications had taken on a project from the large end. The station has always boasted top-of-the-range technology. The proprietor spares no expense in kitting the station out and keeping it up to date.

From a lone station on the outskirts of Lagos, DAAR Communications now operates from nine locations around the nation —Lagos, Abuja, Enugu, Kano, Kaduna, Jos, Port Harcourt, Yenogoa and Agenebode (DAAR Prospectus, 2008: 49). With such a geographic spread, AIT has a virtual national presence and is able to offer an alternative to the NTA network. This structure is unique for a privately owned enterprise and is evidence of a reversal of fortunes for some minority ethnic groups, especially those in the South-South, who have become more visible. The choice of Agenebode (in Edo state) as a site for operations is sentimental (nay, parochial) as it has little more to qualify it than the fact that it is the hometown of the proprietor. It may not be the best business decision, but this is good news for the Etsako people, who had hitherto lacked such representation in the media

Though it does not yet transmit from the following locations, AIT maintains a presence in Yola, Maiduguri, Katsina and Osogbo. The plan is to have coverage in as many as 22 of the 36 states. This can only strengthen the network of stations spread across the six geopolitical zones of Nigeria, while providing an alternative voice at the grassroots. In February 2008, it received approval from the President to start network transmission, though some contend that it had already been operating some form of network transmission[3]. Compared to the NTA stations, AIT does not have standard studios in its bureaux. In Jos, Kaduna and Kano, it has rented space in tall buildings on which it rigs its masts, and with a modest transmitter it is able to disseminate its signals. These have been described by some as glorified offices, but all AIT needs is a small studio for the bureau report, since Lagos and Abuja supply the main feed. This saves on capital investment. It is not known if there are plans for permanent structures in those places. In the meantime, the modest facilities suffice in the penetration of a wider scope of the market. Whether these can stand up to professional standards of television practice is another matter. This is the face of the democratised television service.

Already the organisation has begun international transmission; to the United States and to the UK and Europe via satellite. By 2008, AIT's investment in digital equipment, a new outside broadcast van, expansion of its Abuja studio complex and an upgrade of its facilities in readiness for the launch of the multi-channel high-definition pay-TV service was visible. Rather than negotiate to be on another organisation's platform, DAAR was investing in its own multi-channel service: an upward vertical integration. Though this is an expensive venture, it gives AIT greater control on its downstream operations—the sort of programmes that can be shown. Through its new platform it seems set to accomplish its mission of reversing the information order. For this to happen, it must attract audiences other than Nigerians in the diaspora. That is a different challenge which should be explored in greater detail, yet its modest achievement in this direction must not be underestimated.

Unlike other visionaries before him, the proprietor of AIT is not a politician who seeks elective office. In this way he can be likened to Rupert Murdoch, who presides over a large media empire and wields political influence without being directly involved in the political frame.

AIT's proprietor, High Chief Raymond Anthony Aleogho Dokpesi, makes no claim to any previous broadcasting experience before DAAR Communications, yet he seems set to prove a point through this venture. This would be consistent with his track record. In his days as a student activist in Poland, and president of the African Students' Union, he acquired a reputation as a Pan-Africanist. He developed this reputation as a pro-African businessman during his years in the maritime business. Along with his mentor, Bamanga Tukur, and business partners Shehu Musa Yar'Adua and Basorun MKO Abiola, he established African Ocean Lines. He ran this private commercial enterprise and succeeded, if only for a brief period, in registering the competence of African entrepreneurs in a field dominated by nationalities from more advanced economies. Perhaps through DAAR he seeks to do the same in broadcasting. But it would be naïve to ignore his political agenda simply because he is not directly connected to an established party or similar organisation. His role in politics prior to his foray into broadcasting had been largely to support his business partners, in the course of their political campaign during the Second Republic. That tour took him around the country It was then that he recognised the depth of ignorance about government activities within the Nigerian society, and the need for private-sector participation in broadcasting if this was to be redressed (Ebuetse, 2007: pg 60). In other words, as did other private entrepreneurs in the broadcasting industry, he recognised opportunities in the market that had been created by the way government ran operations. It is therefore obvious that Dokpesi is also politically minded.

AIT's vision is consistent with the threefold Reithian mission for broadcasting – to Educate, Inform and Entertain. It attempts to meet the interest of a diverse audience and offers a 24-hour television service. This requires shrewd management of its schedule; by careful interpretation of how primetime is deployed, the station is able to cater for a wide range of audiences. There is an understanding of which television forms appeal to which type of audience. According to the director of Programmes (interview conducted July 2008), these are carefully balanced to give the station a broad-based appeal. For women, there are drama productions and discussion programmes. It is assumed that men come to the station for the news, politics and sports.. Music shows and drama are the bait for children and young people.

By 1996, when its operations commenced, Nigerian audiences were more critical and more demanding than previous generations. The proprietor recognises that competition in this era, especially with AIT's incursion into international service, would require good ideas and technical competence. Besides investing in the equipment to make this happen, the right staff must be identified. The station is inclined to rely on the younger, more technically savvy generation. Old hands who do not keep up with technology should be prepared to be rendered redundant by those who are conversant with the information and communication technology (ICT). It has no qualms about this. The executive chairman is of the view that the ICT generation may be inexperienced, but make up in their creativity what they lack in professional conventions. Such a position may explain why some within the industry regard the station as having maverick tendencies. But the compromise on professional experience does not solve the problem of staffing. Most institutions lack the facilities to offer industry-standard vocational training, so generally fresh graduates do not meet the mark. This calls for creative solutions from the organisation. Like other stations, it relies on in-house mentoring and local staff training through local institutions, universities and the NTA

TV College. But it also imports expertise from abroad when necessary (it flew in the team of engineers who installed the studio and transmitter facilities for the multi-channel operations).

With the provision of up-to-date equipment, evidence of commitment to investment and the promise of better pay (in an industry which till then had been dominated by government, with its reputation for meagre and irregular pay on some stations), AIT was an attractive destination for broadcasters, especially those in the southwest axis. Staff in the core business areas, Programmes, News and Current Affairs and Marketing, came from NTA, Ogun State Television (OGTV), even the new Channels television. These recruits were relatively younger, and, being part of a new generation, they had some degree of relevant training, education and exposure. AIT rode on the goodwill and fortunes of its sister establishment, Raypower FM, with its humour, good music and likeable personalities. Those who made their mark on radio would be given an opportunity to shine on TV. Both stations were well received by artistes and audience. Even considering the existing buzz, AIT brought a fresh dimension to the industry.

The station is aware of its contribution to national aspirations, in terms of the enhancement of the national image, and of the drive to position Nigeria to attract economic investment. Its mission, as captured in its slogan, found further expression in small gestures such as dress and name codes; presenters who had on other stations used English names took on African names at AIT. Their dress also reflected the local cultures. It seemed far-fetched at the time but AIT nursed the ambition of helping to reverse the tide in the one-way flow of information from abroad, through international transmission. It would begin by having a strong emphasis on local input in its domestic service, and then, perhaps with time, the best of such programmes would eventually be refined and could be transmitted abroad. Such aspirations were an inadvertent indictment of the NTA, which was the national station.

DAAR Communications was committed to identifying and supporting local talent, yet it was also a conduit for the latest on the international music scene. Among its popular early programmes were music shows,, such as *AIT Lunchbreak and AIT Jamz*, and the musical interludes. *Soul Shuttle,* anchored by Namure Edomoiya, designed to suit the late-night belt, was a phone-in show in which people talked about and received counsel for their problems. This musical request show with a twist justified the foray into 24-hour transmission.

At inception, AIT invested heavily in local drama productions and soap operas, collaborating with independent producers. Its local drama productions were the "bankers" for the schedule. Among its early successful drama productions are *Palace, Family Circle, Everyday People, Four's Company and Candlelight*. The focus on family and personal relationships is typical of the models that had worked on other stations. It also reflects the perennial concern regarding television and the erosion of sound family values (Didigu, 2002). The difference in AIT's initiative was perhaps in the sleekness of the productions – the colour and the grand locations. In recent times, it has also introduced foreign soap operas, which proved to be popular. These programmes have consistent slots on the schedule, helping to cultivate the loyalty of women and children in particular.

AIT's aim to promote good-quality, entertaining local content was expressed in a variety of ways and a broad definition of its social responsibility. A range of discussion programmes focus on business, politics and other topical affairs. These audience-participation programmes have been facilitated by the availability of new technology, such as audience access to mobile phones (for calls and text messages). The station had several popular programmes designed to facilitate such communication, especially the panel discussion shows such as *Kakaki* and *Focus Nigeria*. Anchored by Tosin Odukoya (later Dokpesi) and Jika Ato, then Kunle Adewale, and Imoni Amarere *Kakaki – The African Voice* is a regular breakfast show. *Focus Nigeria*, anchored by Gbenga Aruleba, also features on the breakfast belt but it is more provocative, adopting an adversarial (fact-finding) posture on topical controversies. Both programmes tend to have a political focus, both have panels of experts or opinion leaders, and they accommodate viewers' contribution through phone calls and messages. There are also independent productions, like *Soni Irabor Live* and *Patito's Gang*, which tend to have discussions and features on political issues as well. These phone-in shows give an indication of the breadth of the fan base and offer ordinary citizens opportunities to express their views.

With these programmes, AIT can claim to have contributed to

the robust nature of the political debates by presenting perspectives other than those on the official channel. Its coverage of Nigeria's Truth and Reconciliation Panel in 1999 at the dawn of the return to democratic government, following the demise of General Sanni Abacha, the notorious military head of state, marked a new beginning in outside broadcasting. The coverage was a media spectacle in itself. The nation has since been treated to live coverage of proceedings in the Houses of Assembly from both AIT and the NTA. This is a new dimension to surveillance of the political class and brings the nation a step closer to fostering accountability.

The station also offers a forum for expression on its Yoruba-language programme *Mini Jojo*, the variety programme featuring Bashiru Adisa and Ambrose Shomide. This studio-based audience-participation programme promotes and debates aspects of the Yoruba culture. The vox pops with men and women on the street are regarded as a winning formula and are used on a range of programmes. According to the executive director, Programmes, Tosin Dokpesi, the station has tapped into the love Nigerians have for expressing their views. This political activism in support of democratic governance since the run-up to the Third Republic mean there is no shortage of issues to discuss.

The *Bisi Olatilo Show* and *Zoom Time* are examples of locally produced syndicated entertainment shows. The *Bisi Olatilo Show* is worth mentioning. It is a version of the social diaries, with highlights of weddings, birthdays, funerals and lavish parties. The programme is a document of private celebrations, typically of public figures such as political and military leaders, captains of industry and other celebrities. It is in this parade of the leaders' lifestyles and expenses that such otherwise innocuous tabloid tales become more than mindless entertainment. The programme also features on other channels. Indeed, other stations, including the NTA, have their own versions of the format.

AIT regards itself as the station people prefer because it knows how to meet their needs. Besides informal feedback transmitted through its staff and their personal contacts in the community, it gets indications of its performance from audience-participation programmes. It uses reports from audience research agencies such as Zus Bureau to assess its performance. Though it operates on private funds, and without any openly declared financial support from any government, federal or state, AIT gives time to public-service announcements, charitable causes and issues of public interest, as is done on government-owned stations.

AIT takes pride in reflecting African culture in up to 90 per cent of its programmes, suggesting that audiences may find better cultural affinity with the station. It sees itself as an ombudsman for people fighting for justice and fairness, especially those who for too long had been without a voice in mainstream politics. This is evident in its championing of the cause of the people of the Niger Delta (the South-South Zone) yet it claims it is non-partisan. AIT's goal is to be a trailblazer; to make sure it is "the station of choice"; that its programmes are preferred by the audience in a very competitive market. Though it is a claim that others in the business contest, AIT prides itself on being "the brain of the industry". According to Imoni Amarere,

> We [AIT] do the thinking for other stations. For example, AIT initiated the idea of network bureaux on Nigerian television [news]. Its picture quality is a reference for other stations. On the whole, it has enhanced the aesthetics of television in Nigeria. The station has been able to do this by the quality of its production and transmission equipment, and its conscious effort to learn from the practices of the world leaders in the business. Even when staff are not travelled, they have access to the practice of others through cable channels.
> (Interview conducted in July 2008)

AIT is not alone in monitoring other stations. It is standard practice to find principal officers monitoring the output of their station, while also keeping an eye on the output of competitors and mentors. Evidence from an observational study in 2008 showed CNN, Al Jazeera and Sky News were the typical foreign channels monitored.[1] Such monitoring is important, as these stations are available to audiences via pay-TV and they have become the yardstick by which viewers judge how up to date local stations are on global events.

In its first 13 years of existence AIT has had its fair share of drama. It started transmision in 1996, just three years after the annulment of the election that was reputed to be the country's freest and fairest. AIT was established in a military regime, though the struggle to restore democracy was rife. The print media in particular had been the "avant garde" for this struggle, while the NTA was discredited for its passive, unsupportive, possible antagonistic role in that struggle. It has been noted that the broadcast media, rather than ". . . stand up for constitutionality in governance and leadership ... actually accepted and cohabited with the soldiers and even descended to the level of declaring them as keys that opened doors, or having heads that the cap fitted" (Salihu 2002:143). That observation referred expressly to jingles run on the NTA in apparent support for a military head of state who sought to transform into a civilian president. So compared to the NTA, AIT's performance was regarded as provocative, possibly mischievous. Perhaps for AIT the struggle for democracy is still on-going, but in maintaining its aggressive adversarial posture (whether through its staff or invited participants, including members of the public) it is regarded as being unduly sensational and is censured in different ways. There was the case of a top government functionary who accosted a reporter because of comments that had been made on his show. "You are the ones who have been embarrassing the government," the reporter was told. The leader was armed with a file in which he had recorded incidents, dates and a list of unsavoury remarks made by callers to the show.

In another incident, the same reporter was told unequivocally that he was not welcome at the president's table at a reception held for the press after a briefing. It was made clear that the reporter could not challenge the top man in public and hope to eat at his table. He had to leave the esteemed company and there was no support from other "gentlemen of the press".

Its radical posture has put AIT on a direct collision course with the establishment on several occasions. The scathing results of such encounters presented the station as the victim; the one willing to suffer for the masses. Twice in its short history, AIT has been shut down; a fate that had hitherto been common in print media, but unknown in broadcasting. The first instance was in 2000, when the facilities were sealed by receivers for about six months. The management was certain that the issue at stake was more than financial. Rather, it reckons that the squeeze from creditors was at the instance of government forces that sought to bring the organisation to its knees. The other closure was at the instance of the regulatory body, the NBC, over what was regarded as a distasteful news report. On the first occasion the quick intervention of the Minister of Information shortened the closure but not before it had drawn sharp criticism from other broadcasters. The second incident described below involves its coverage of an air crash. Its distinct approach to this task is an example of its populist approach, which makes it the tabloid of Nigerian television; a charge made by the regulators and other broadcasters. AIT got its knuckles rapped for that coverage, even though it remains proud of the role it played in what was a national emergency. This exemplifies the controversy associated with different expressions of social responsibility. Some attribute the variation in perspectives to the drive for larger audiences that is characteristic of a commercially competitive service.

On Saturday 22ⁿᵈ October 2005, a plane (Bellview Airline Flight 210) crashed. For several hours there were conflicting reports in the media about the situation - the location of the crash, number of fatalities and the number of survivors if any. It was obviously an anxious time in the nation. People wanted to know what was going on; in the era of 24-hour television, news was meant to be immediate, believable and visual. At least this was the reasoning at AIT, which was the first to report the site of the crash. Though AIT's version clarified the situation, it also heightened the dramatic value of the story and aroused the emotional response of the audience. The account was similar to that which remains on the CNN website.

> Dismembered and burned body parts, fuselage fragment and engine parts were strewn over an area the size of a football field near the village of Lissa, about 30km (20 miles) north of Lagos . . .
> (CNN International.com 117 Killed in Nigerian Plane Crash, Sunday 23ʳᵈ October 2005 http://edition.cnn.com/2005/WORLD/africa/10/23/nigeria.plane/ Last accessed 10ᵗʰ April 2009)

Many professionals were offended by AIT's treatment of the tragedy; they felt that it exploited the situation, deepening the trauma of the relatives of the deceased. The following account of the incident by the BBC is telling.

> The plane was first reported found on Sunday morning by a police

helicopter search team near the rural town of Kishi, Oyo state, 400 km (320 miles) from Lagos. It was suggested 50 people might have survived.

But officials later retracted statements about the plane's location and survivors after a TV crew said it had found the aircraft near the village of Lissa in Ogun state, about 50 km (30 miles) from Lagos.

Images of mangled bodies, twisted chunks of metal and ripped luggage were broadcast.

BBC News Africa, 23rd October 2005 All Killed in Nigeria Plane Crash http://news.bbc.co.uk/2/hi/africa/4368516.stm last accessed April 10th 2009

In this, AIT was regarded as breaching conventions on the consideration for the relatives of the deceased, and the regulatory guidelines on use of graphic details during disasters (NBC Code Sections 5.5.5 & 5.1.15). The NTA claims that one of its reporters had found the site much earlier than its reports suggest. Danjuma Abdullahi, who had been on a naval plane on a separate assignment with the Nigerian Navy, had sighted the crash and recorded visuals. However, there was no rush to report the story, since it was NTA policy to exercise restraint in such a matter, to allow time for relatives to be informed officially. The NTA's version of the story was "clean" and devoid of the graphic details in its competitor's account. The NTA refrained from using any visual reports showing mutilated limbs, disembodied heads or other body parts. In its view, the use of the images of the body parts, even if more dramatic, was distasteful because victims could be recognised by unprepared close relations. In this instance, the NTA had other concerns besides the desire to be the first to report; being responsible was more important than being competitive in a national tragedy. In the end the NTA is proud that Abdullahi's pictures were preferred by the world press over those of its other rivals.

The NTA's definition of its role predisposes it to the accusation of it being sycophants overly dependent on government patronage, but AIT's orientation has brought it the charge of being reckless. The controversy is captured in a newsman's appraisal of television in Nigeria.

A lot of the commercial stations are irresponsible. The dictates of commercialism encroach on decency, ethics and balance . . .It is not good in the media. We should try, in as much as we want to keep our stations on air, to be very responsible in the media; that we do not . . .damage. . .the very fabric of the society in which we operate. I believe that some people may come to the public stations to air their views and they will be refused, because these are not views that are in the best interest of society at large, but those views will find their way into private stations. These stations agree to air these views because in doing so, they are able to present views that some in society want to hear, though they are not views that need to be aired. The stations make money from adopting this popular appeal. [People in] broadcasting should pursue decent money, not money got off the blood of the people. [Government stations are] not being lame ducks when certain stories [are not taken], it is because we [public broadcasters] know the precipice [that] situation[s] will lead to. (Ladan Salihu ex-NTA News; zonal director, Federal Radio Corporation FRCN –Interview conducted July 2008

In such situations, the National Broadcasting Commission has a duty to adjudicate on behalf of the audience. This accounts for the apparent antagonism between the regulator and broadcasters. The NBC intervention is often regarded as high-handed, as was the case in this situation. The regulator is considered out of touch; its rules are seen as strict and outdated. Its attention to issues of decency which bothers some at AIT includes detailed attention to presenters' dressing; except when presenting sports programmes, they cannot wear T-shirts. There are concerns about sets, complaints about backdrops and about music clips which are regarded as vulgar. The news report of the plane crash was yet another instance. The station presents itself as persecuted by the authorities, presenting its case in the court of public opinion, where it finds support. It is apparent from this that there are different standards of decency competing for supremacy.

Often debate about decency is evidence of a class struggle— whose taste should be accepted as the norm? From all indications,

AIT, with its populist appeal, has contributed to the "tabloidisation" of television news. If this is an expression of a cultural struggle, the gains of democratisation are apparent. In this context tabloidisation refers to the tendency to package news stories for broader appeal, often leading to sensationalised accounts that generate attention among those who otherwise might be apathetic about the issues being discussed. Tabloidisation on television news is, however, not limited to Nigeria; it is part of a global trend, hence the view that regulators are out of date. Winston (2002) presents a review of the key arguments and the evidence of this trend in Britain. The issue is more than style; it cuts to the heart of the efficiency with which the medium performs its role—what types of issues are covered, how well are these understood, and by whom? In other words, who gets to be involved in this space where rational discussions need to take place? For this reason, healthy debates between practitioners and regulators are desirable.

AIT has good cause to feel paranoid. Apart from these two incidents detailed above, it has also been gutted by fire a number of times, resulting in its being off the air for varying periods. If good can come from such situations, then perhaps it is the fact that the organisation was able to judge the intensity of public feeling in its favour. There was a practical demonstration of support, from viewers, wellwishers and prospective investors at home and abroad after one such incident. From a fund-raising appeal for the station organised by such supporters, 1.6 million naira was raised. This was a paltry sum compared to the 350 million naira that the station needed but if the public was prepared to donate this amount without any expectation of returns, then it could be that broadcasting was ready to be organised as a limited liability company. It took the decision to raise money from the Stock Exchange and executed this in February/March 2008. The response to the subscription was overwhelming. The venture had been well promoted through the media, especially AIT and Raypower, also owned by DAAR Communications. So popular was the scheme with the public that special measures had to be taken to avert over-subscription. This may be proof of a market that is ready to invest in its broadcast media. This is clearly a new era; a new interpretation of the fact that airwaves actually belong to the audience, and that broadcasters merely hold the authority to run the stations in trust. It also confirms the viability of another stream of funding that had never been explored for television broadcasting in Nigeria. AIT is the first television station owned by a publicly listed company.

Like other Nigerian television organisations, AIT has a website that can be developed to take better advantage of recent developments in communication technologies. AIT is forging new paths and courting a new generation of viewers, especially with its USA and UK service. Courtesy of these audiences it gets mentioned on the social networking sites—Facebook, MySpace and even Twitter. This is great for fostering the integration of Nigerians, including those living abroad, but AIT has hardly caught the eye of Western audiences. Yet, by its special features from the African continent, such as live coverage of the inauguration of the Ghanaian President John Atta Mills from Accra on the 7th of January 2009, AIT is striving to cultivate a following among other Africans in the diaspora. Here was a Nigerian station presenting a Ghanaian event to global audiences. It launched its own satellite service, DAARSAT, in October 2008, and direct-to-home pay-TV service. There were plans to manufacture decoders in Nigeria to boost technological transfer and make the service more affordable locally. As many as 40 channels from different Western and a few African countries are on offer. A broad mix of news, entertainment, music, sport, religious and lifestyle channels, including two in indigenous Nigerian languages, Yoruba and Hausa, will be offered. Laudable as the effort is, it provokes two anxieties regarding the loss of control that exists on a terrestrial operation. Succumbing to the irresistible pull of Hollywood and other Western fare makes economic sense, even if it contradicts what the station stood for at its inception. In more direct and immediate ways than was known on free-to-air terrestrial television, subscribers demonstrate audience power to influence programming on the digital satellite television. The intensive financial commitment, along with the political manoeuvrings required to survive in a competitive market, may compromise the integrity of organisations with such high ambitions

From its record, AIT has demonstrated the utilitarian value of technology. Going by the mission statement, the vision and programming, this organisation seems to have found a happy middle ground between private entrepreneurship and public service; the prominence of political programmes on its schedule appears to be an expression of its social responsibility. Yet the fact that AIT is not making a profit is incongruous with commercial enterprise. What, then, is the real gain or motivation for the venture? A sceptical view suggests that attention to political issues is merely for the economic sense it makes (hence the populist style); it is a venture pursued just because it brings in the audiences. To the more cynical observer, television business, as with other media businesses, helps generate goodwill and political influence. It is too powerful a tool to be neutral, especially in the hands of influential men who have friends in the corridors of power.

Community Television Service

Previous sections of this chapter discussed the inception of private broadcasting occasioned by global and local deregulation. But these do not account for the phenomenal growth in television broadcasting initiatives. The NTA has been pivotal in the expansion that occurred in this third phase. Nigeria boasts Africa's largest television network. In December 1985 Yemi Farounbi (then general manager of NTA Ibadan) wrote that television had come of age in Nigeria. ". . it has grown from one station in 1959 to over thirty stations, from two transmitters to over fifty transmitters, from a few television sets to over four million sets; and from a few hundred urban viewers to over twenty million potential viewers." (Teleview, 1985: 3)

By January 1991 (just before the third wave) NTA had 24 production centres and 56 operational transmitters; it was responsible for broadcasting 85,000 hours of programmes annually. In addition, there were 14 state government-owned stations. To ensure that national coverage is adequate, there is an NTA station in every state capital. However, considering how vast the nation is, these stations could not guarantee adequate coverage at the grassroots, as has been discussed. Consequently, the idea of community stations to complement the service was conceived. In this third era, television has evolved. While NTA International Service has been launched to reach audiences beyond Nigeria, the NTA has sought to consolidate national coverage. In 2008 the authority had nine network uplink centres and 101 stations, about half of which are community stations (some of these are not yet fully functional). The NTA has an estimated 90 million viewers.

This means slightly more than half of the Nigerian population now views television. This contrasts sharply with the picture painted by the 1996 aggregate for African households (Tomaselli & Heuva, 2004: 96). That figure was put at 3.5 per cent for television-set ownership, thus suggesting about 21 per cent viewership rate if each set attracts six regular viewers. Clearly Nigeria is far ahead, possibly because of the attempts to bring stations closer to the audience. That is why the operations of one such community station will be examined in this section.

NTA Zaria is an example of a community television service; its audiences are spread across nine local governments. It is located in Kaduna state, home of RKTV (later NTA Kaduna). Zaria is about 70km (less than 45 miles) from Kaduna. Before it became a full fledged station, it was merely a relay station for network signals. Apart from being a traditional centre, the headquarters of one of the original Hausa states (Zazaau), Zaria is a university city. It is the home of the Ahmadu Bello University, the Aviation School and a military school. In spite of the largely rural population in its immediate vicinity, it has an urban population with the diverse mix of ethnicities. This description is typical of most of the community stations established by the NTA.

There are community centres in places like Ile Ife in Osun state. This is regarded as the ancestral base of the Yorubas. It is also home to one of the first generation of universities in Nigeria, Obafemi Awolowo Univeristy. Oyo is another historical town and significant cultural centre served by a community station. These places had been served by the premier television station Western Nigerian Television Service (WNTV), later NTA Ibadan. Other strategic cultural or commercial centres that were in the old WNTV domain but now have community stations include Ijebu Ode, Imeko, Ogbomosho and Saki in the South West; and, in the MidWest area, Sapele and Warri.

There are community stations in the South East in Onitsha, Awka and Okigwe, and in the South-South in Ogoja, Eket, Ikom and Brass. Communities in the Confluence area of the Rivers Niger and Benue (in Niger and Kogi states) also have stations, for example in Bida, Lokoja, Ayingba, Kabba, Kaima and Keffi. Community stations around the Middle Belt and up to the northern states include Ankpa, Birnin Gwari, Birnin Kebbi, Biu, Dutse, Gusau, Ganye, Lafia, Pategi, Kontangora, Oturpko, Takum, Numan and many more. From this array of locations, it is clear that there is an attempt to have a national spread and to improve the penetration of messages to the grassroots. Many of these communities, like Kaima in Kwara state or Ogoja in Cross River state, were so far from the state capitals that without such interventions, they were in effect marginalised within the state. In some cases community television stations had been sited in communities close to each other. This may be indicative of the distinct outlooks (identity and aspirations) of the people. Having distinct services in such communities may thus help to reduce communal conflict. In any case, with effective television service that feeds into the national network when necessary, television may finally be meeting the challenge of fostering dialogue and national cohesion.

This project was realised within a democratic dispensation and must be viewed as a dividend of democracy. Yet one should not rule out the chances that the choice of some locations was politicised, despite the logical justification that can be found. This would not matter if the stations were then properly funded and well managed in the interest of the people. That stations are in closer proximity to the people should make it easier for the people to feel a sense of ownership of the station and its programmes. Perhaps then the carefully planned messages can have greater impact. It will be even better if they can be involved in identifying common problems and finding the solutions. Perhaps then television can be more relevant to plans for community development. For the vision to be actualised, the people must also be ready to embrace the medium and support its existence. This was well illustrated in the example in Zaria.

The Case of NTA Zaria

A visitor's view
The station is on the outskirts of Zaria in a purpose-built complex. The mast is on the premises, so it helps to identify the location. Visitors to the area must take care not to be confused by the nearby telephone mast. The area is quite tranquil, as it is close to not much else besides the State Water Board. This is a remote location for a

community station, as access to the station would require deliberate effort. Yet people do appreciate the station enough to make the effort. NTA Zaria is a station that exists against all odds.

The staff strength is very lean; the premises serene. There were no people milling around, there was no light either. Next to the building was the generator house that is a usual sight in Nigeria. The generator was resting for some time, but not for long. Inside the building were a few offices, nothing fancy, but they were functional. In one of these was the officer in charge. At the community stations there are fewer people to perform a range of duties—they were already multi-tasking while staff at regular stations were still trying to get their heads round the policy.

A few men rode in to the premises on their motorcycles, first one, then another, till the group was complete; a local musical band had assembled. They must have been there to keep an appointment. Like clockwork, the generator was cranked to life—someone said something about AM transmission. Was it time to take a network feed? The band was ushered into the studio for their recording. The station was getting the most from the generator.

The team in Zaria had a very positive orientation to the mission of community broadcasting. It is perhaps the reason why they could cope. According to the management, people are working beyond the call of duty to make the station viable.

The transmission day is short because of the paucity of power supply. As with other stations in the nation, NTA Zaria relies largely on privately generated electricity. The generator is on for six to 12 hours a day. It is used for transmission, which runs in two shifts. Morning transmission begins from 6am till noon. There is a break till 4pm, when the station resumes for evening transmission till midnight. This is usually run on a diesel-powered generator.

There is no back-up for the generator, since what should be the emergency supply has become the main source of power for the station. The transmission log showed that the generator was on for an average of 42 hours per week, compared to seven hours of public power supply. For every hour of public electricity, the station relied on the generator for six hours. That is an immense challenge for a station that is required to be self-sustaining; NTA commercialisation policy requires all its stations to be self-sustaining. How does a station like this sustain itself? Which advertisers would be interested in such a market?

The station's records show that it is patronised by local politicians, local businesses (for example, Hon Mustapha Bawa), religious and educational institutions, (Makaratun Allo, Dakachi) groups of migrant communities residing in Zaria (for example, indigenes of Akwa Ibom and other associations who have "town meetings"), charitable organisations and representatives of the donor community. These all play a part in keeping this station alive with their paid announcements and advertisements. The Soba (Gimba) local government also places adverts.

NTA Zaria is an example of how the rural economy can be tapped to sustain television business. One just needs to find those with an interest in the community; those who have the potential for growing their business concerns in the area. The Zaria example showed that at times, one needs to break from convention and find the appropriate and acceptable contractual terms. In Zaria there was evidence of bartering. Because diesel was so precious, the station agreed to exchange airtime and service for supply of diesel from a private educational institution and a local government authority in the area. This arrangement was clearly of mutual benefit, as was explained by the station manager.

> Nuhu Bamali Polytechnic gives us a drum of diesel for a month. Their students come [to the station] for industrial attachment . . . This is a community station and should be subsidised by the community. It is a joint venture, between the authority [the NTA] and the community, and the community has not let it down. Since we started we have never stopped for a day because there is no diesel. (Mohammed Kabiru, Station Manager NTA Zaria interviewed July 2008)

In the transaction mentioned above, the station derived two benefits —labour from the students whom it helped to train, and the diesel. It also generated goodwill. This station did generate revenue, but its expenses on diesel proved to be a great drain; without that added expense, it might have had more to invest on making programmes within the community.

Fig 5.4: NTA Zaria premises

Fig 5.5: Emir of Zazzau, with NTA Zaria Station Manager and Zonal Directors

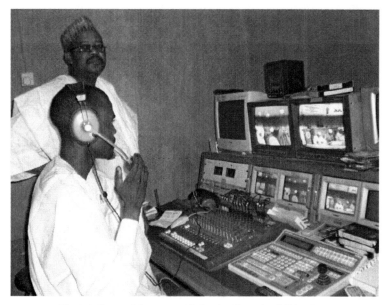

Fig 5.6: Station Manager inside Studio Control at NTA Zaria

Community support for the station had been rooted in the traditional institution from its inception. The Emir of Zazzau (the traditional ruler of Zaria province) has been supportive in practical ways. He donated a digital camera to the station and often intervened with the Electricity and Water Board on the station's behalf. The station manager reported that the Emir had intervened on occasions when planned power outages (load shedding) by the electricity supplier became excessively disruptive. This model of television service shows that television is everybody's business. There are lessons to be learnt from this. The first is that the station manager needs to know the business of television and have a professional outlook. He or she should also know the community, the people - their information and development needs, their industry, and creative ways of extracting resources from these. In this instance, the officer in charge is an indigene who understands the social systems and how to make these systems work for the station. He knows the people whose support is vital, and maintaining contact with them constitutes a central part of his job. He also knows how to address these people. The station's manager is the advocate for the station, marketing its programmes, scouting for programme ideas and talent to be produced; acting as a reporter, ensuring that no important lead is missed. To be all these, the station manager must appreciate professional processes within the business.

Programming on NTA Zaria is a mix of local and network programmes. Reliance on network programmes reduces the burden on the local station. The morning belt, for example, is dominated by the network relay of *A.M. Express*, which takes feeds from Zonal Centres around the country. *Road Safety Watch* and *Human Rights Platforms* are examples of network programmes on the early evening belt. Some of these are independent productions that seek national reach. They are informative (if not instructional) in their orientation.

Local programmes tend to be basic, featuring local amateur artistes. For this reason the station can get away with offering their artistes modest fees. In this regard, operations are similar to the situation that obtained in the earliest stations. This is how the station has coped on its meagre budget. The local fare includes dramas and comedies to help people unwind after a hard day's work. These include *Shiga Sharo ba shanu* and *Gida kase o*. There are also musicals, interviews and discussions through which the cultural heritage is transmitted. For the children, there are debates and quizzes sponsored by local politicians. Local schools are featured on magazine programmes free. The family-oriented programmes are designed to reflect the culture. At times they are used to challenge retrogressive attitudes or promote changes in established harmful practices. Programmes have been used to explore links between early (girl-child) marriages on the incidence of vesico vagina fistula (VVF), which is caused by prolonged labour in childbirth, often when the mother's body is underdeveloped, because she is malnourished or underaged. The incidence of VVF is also linked with female genital mutilation and contributes to the exclusion and subordination of some women. Some programmes adopt formats that facilitate negotiations on the recognised duties of women in the home and acceptable strategies for enhancing the quality of life in the family; others consider how to address common social problems such as involvement of children in street trading.

Some programme concepts are a result of suggestions from members of the community, some arise from feedback on existing productions. This way, programme producers harness the particular interests of their audiences. Even without a formal feedback mechanism such as the community viewing clubs, the station does not lack suggestions from its viewers. It merely helps to adapt these to formats suitable for the medium. Such collaboration is made possible by the proximity of the station to the people it serves.

There are public enlightenment efforts discussing preventable diseases on programmes such as *Health is Wealth* and *See your Doctor* and their Hausa versions, *Donyalinku* and *Lafiya Uwurujinku*. They also alert the public to issues affecting their health delivery services. The station helps to manage information from government and the community. It promotes self-help but also champions the cause of the people in ventures that require government initiatives (such as road construction or clearing blocked drainage systems). In this regard it is similar to the stations in the state capital, but as these are situated in the rural areas, surveillance occurs at the grassroots, in areas that might otherwise be neglected. Cameras are taken to such sites and the attention of the local government authority is called to these problems in the full glare of all. This forces the hand of those in power. By so doing, television is the audible and visible voice of the people and returns power to them.

NTA Zaria also devised programmes to promote accountability in governance. A programme like *Bako Mu yau* brings politicians face to face with their constituents. However, the fact that the station is so dependent on the political class, along with other members of the elite, and that this is a sponsored programme, raises questions about its ability to retain editorial control in the interests of the audience. The effort must, however, be commended for the attempt to conduct conversations in the community. There are other examples of community-spirited programmes. On one of these, the station helped to raise funds for less privileged members of the community, to pay hospital bills for indigent patients who would otherwise languish or die from their ailment. It has helped to avert the ethnic hostilities that have occurred in other parts of the nation. This is done in collaboration with the paramount traditional ruler in the area. In one case before elections, it arranged for the Emir to address the viewers. With the immediacy and visibility of television reports and the respect that people have for the Emir, this was a winning strategy to avert political trouble. Besides, with television coverage, people are able to keep up with unfolding events, be rational about them and remain calm. These examples show what television can achieve when handled responsibly; but the onus is on media operatives and the politicians and community leaders to decide how the medium is used. In the event that any one of these exploits the privileges offered by the medium for their selfish ends, television may fail in its critical duties in society. Thus, for community television to be successful, it requires altruistic leaders who are committed to the service of the people. There is another critical lesson to be drawn from the running of this community station. The success of its interventions depends on the integrity of the political structures with which the station is inextricably connected. Where traditional rulers have been incriminated in local or national political struggles, the base of acceptance required by community television programmes may be compromised. This limits the prospects of community television in conflict-ridden areas where community leaders are held in contempt. Similarly, where television narrowly defines its roles, when the station is content to report rather than be involved in averting conflicts, the collaboration between the station and the community could be less meaningful.

Community television has the potential to increase the penetration of television into households. On one level this is technical; depending on the location of transmitters—an issue that engineers have battled with since the birth of television in Nigeria. Weak signals, interference from telephone masts and poor electricity supply have increased the scepticism about this prospect. These problems require a range of engineering solutions. From the experiences over the years, there is a better spread of the transmitters. On a good day, the signals from NTA Zaria can be received in other parts of Kaduna state, and as far away as states bordering it—Niger, Kano, Katsina. But the reach of the signals does not guarantee audience reach. In this it becomes apparent that the potential of community television relies on a social response as well as an engineering solution. In the past the costs of configuring the engineering solution and social response culminated in the view that television is an expensive medium. The cost is even higher when the financial implication of staffing so many community stations is considered. Yet from this example comes a new corps of station managers who have a more optimistic view of the medium.

To them, TV is not as expensive, as previously thought,

> . . . now that there is light in villages, and small generators are affordable, people are able to see [what is on television]. Advertising is better on television because you can show the product, not merely describing it. At times you run the phone numbers that people can call for their orders. Television is a better social diary, showing weddings and social events. Religious seasons and activities offer opportunities for the leaders to share messages of goodwill and identify with their communities.
> (Administrative officer, NTA Zaria interviewed July 2008)

What is left is to find the programmes and the goodwill to support the initiatives that bring television closer to the people. From all indications, this is not lacking across the nation where community stations have been established. The following is an account of how another community helped to establish its station. It is the story of NTA Kabba.

> In Kabba [the NTA] gave the station a one-kilowatt transmitter. When the building [for the station] was at roofing level, there was a change in government. The new government [ie, administration in the state] stopped releasing money for the project. People in the community were levied; this was spearheaded by the [local] Christian Association of Nigeria. The building was completed. The generator for the station was supplied by the community, and the [traditional ruler] the Obaro of Kabba, Oba Michael Olobayo, provided accommodation and transport for the first general manager."
> (Jimmy Atte, executive director, Programmes, NTA Interview conducted July 2008)

This commitment from the community shows its desire for a television station. People's sacrifice is laudable and reminiscent of that found in the 1960s. This must be sustained for grassroots television to succeed, as revealed in the Zaria example. However, recent developments raise new issues with the incursion of privately funded initiatives into this sector of broadcasting. Confluence TV (CTV Lokoja 55 UHF) was established as part of Confluence Cable Network, owner of Grace 95.5 FM, in July 2008.

The station is owned by an influential member of this community who is also a politician, Senator Tunde Ogbeha. This is further evidence that television is still regarded as a critical tool for governance and development. Confluence TV is expected to foster development in Kogi and its environs; communities in neighbouring Niger, Kwara and Benue states. Parochial sentiments apart, there is justification for the station in the principle that people are entitled to reasonable television service, irrespective of where they reside. Its close proximity to the people means it can offer a more relevant service—prompt reporting of local events and information that will contribute to the uplifting of the economic and social wellbeing of the people.

Confluence TV's designated market is largely rural, and communities that are often ignored. However, the chances of reaching audiences in Minna, Ilorin Makurdi and Abuja could make it more commercially viable. Yet appealing to audiences so far flung could compromise attention to the primary targets in the Lokoja area. Perhaps it could synergise with a radio service as other stations have done in the past. It is not clear that it can rely on communally generated funds as government-owned stations do, even though Africa Independent Television has shown that private television organisations can enjoy public support if they adopt a populist approach. In any case, as has been discussed, stations need to be creative in identifying streams of funding. With this privately owned community station a new business model for television operations may soon be developed in Nigeria. The challenge is to find one that will not compromise the mission. In the current regulatory arrangement, licences for private community station are so expensive that the commercial drive becomes an imperative. This is why some broadcasters such as Marc Emakpore[5] (interviewed in July 2008) believe that the community licences should be free. He advocates that stations be supported by subsidies taken from the commercial broadcasters in the hope that community stations can be relatively independent and able to pursue development objectives unencumbered. The stations would be protected from the commercial and political stranglehold. That, he said, was the initial proposal to the National Broadcasting Commission prior to the compromises necessitated by political and economic considerations.

Figs 5.7: Local performers in the studio - NTA Zaria   *Fig 5.8 Transmission mast at NTA Zaria*

Fig 5.9: Community representatives at opening of NTA Zaria
Images courtesy of NTA Zaria

On the whole, stations based in more cosmopolitan metropolises appear to be at an advantage over rural counterparts because advertising revenue is more easily sourced from industrialised areas. Yet some people, like one of Nigeria's first television producers, Chief Segun Olusola, are convinced that with a little resourcefulness, managers of community stations can find enough support from provincial economic and cultural structures to make their outfits viable.

The examples from Zaria and Kabba indicate the merit in this position. In a personal interview (July 2008) Olusola argued that communities are regularly organised for events and projects, they pool resources to celebrate a range of festivals, and they can do the same for community television stations. Such community involvement may go to prove that community television stations have a greater advantage, not in terms of revenue generation and provision of support infrastructure but in their ability to generate relevant programme ideas, in audience affinity with the station and in the positive response to messages. If it works, this will be the utopia that development communication has long sought.

Notes

1   The incident happened on 16th September 2008; Channels TV premises in Lagos and Abuja were raided and later shut down by men of the State Security Service. Top officers, including the general manager and the controller, News, were arrested for a report that the President of Nigeria might resign from office on health grounds. His poor state of health was common knowledge, but more importantly, the story—credited to the News Agency of Nigeria (NAN)—had also been reported by Agence France Press. The NAN report was apparently a hoax, but Channels' operations were suspended for a few days.

2.   TV Africa is a transnational media organisation that seeks to collaborate with terrestrial broadcasters, offering them programmes and selling their audiences to media buyers. Like Multichoice, it began its operations in South Africa and supplies a range of entertainment and sports programmes. However, it lacks the direct control enjoyed by Multichoice in distributing its fare via satellite (see Mytton, Teer-Tomaselli and Tudesq, 2005).

3.   Because AIT was operating from multiple locations, it was possible to transmit signals from these simultaneously, thus constituting a network long before it was licensed to do this.

4.   This refers to the study tour of a sample of stations across Nigeria that I conducted in July 2008.

5.   Marc Emakpore is a television broadcaster, who has worked for the Nigerian Television Authority and was a founding member of staff of the regulatory agency, the National Broadcasting Commission.

Chapter 6

Organisational Structures in Nigerian Television Stations

The business of television is necessarily teamwork. This is reflected in the management structure. There are slight variations in the job titles and the way different organisations assign duties but the essence of the business remains the same. To this end, the industry is still rather homogenous. This chapter presents the management structure of the Nigerian Television Authority (NTA) mindful of its 1991 and 2008 incarnation. This structure which evolved from what obtained in Western Nigeria Television (WNTV) remains a reference point for all television. This chapter is not overly concerned with details of changes but will highlight patterns (contrasts and areas of consistency, especially since 1990) and lessons to be learnt regarding core areas of operations.

The management structure in television stations typically consists of six key operational areas. These are Administration, Finance, Marketing, Engineering, Programmes (along with Production Services) and News (and Current Affairs). This is the case whether the organisation operates at the community, state or national level, regardless of who owns it. By 2008, NTA had evolved as the largest television organisation with seven directorates. Finance and Administration was one of these, Directorate of Training and Capacity Building had been established. The Director-General's office was also regarded as a separate directorate. This structure reflects the organisation's priorities and aspirations.

The chief executive officer at television stations is usually the General Manager. Whereas General Managers were once of the assistant director cadre, in 2007 officers of the deputy director cadre were deployed to serve as General Manager to streng-then management capacity. The NTA alone has 101 stations in state capitals and local government areas. Nine of the stations are network centres headed by Zonal Directors. Network centre status has been accorded to the pioneering stations in Ibadan, Enugu, Kaduna, Lagos, Benin Makurdi (and Jos). Other network centres are Maiduguri, Sokoto and Port Harcourt. This distribution reflects the national geopolitical structure and should make for effective coordination of the NTA stations. The authority's headquarters, was initially in Lagos, but it has been moved to Abuja, the Federal Capital Territory and seat of federal government. At the helm of affairs nationally is the Director General, whose appointment is political.

The Director General is the chief executive officer. He is assisted by a team of Executive Directors who oversee the operational directorates. This team makes up the management board, which sees to the smooth running of the organisation. Executive Directors are assigned to oversee directorates according to their professional competencies. There are some whose experiences enable them to cross over between different directorates, especially between News and Programmes. At this level of management they will have acquired sufficient experience over the years of working across departmental lines. Any one of them is qualified to be the chief executive. Each Executive Director has jurisdiction on his or her turf, but a Director General as chief executive has to cater for the different professional concerns involved in the business.

The task of maintaining links between a television organisation and the diverse parties with interests in it falls on the chief executive; in the case of NTA, it is the Director General. At the station level it will be the General Manager. At private stations such as Africa Independent Television and Channels, where the chairmen of the organisations are directly involved in the operations, they perform the function of the Director General. The most obvious party with an interest in the TV business is the proprietor. For NTA, the government is regarded as the owner, since it was established by the government, which continues to bear part of its costs. This view sparks some controversy—how should the government be defined if it is deemed to be the owner of television? There is a view that the station is owned by the people—the tax-paying public whose contributions keep government machinery ticking over. This assumes that the viewer's interest is paramount according to the libertarian theory. The viewer in this case is regarded both as a consumer and a citizen, and the industry is subject to market forces. Another view suggests that television is subject to ideological control. This paternalistic (authoritarian) position holds that the medium should be accountable to government functionaries (not civil servants) because they rule on behalf of the people. People who run the affairs of state are custodians of the people's will. Strictly speaking, such an argument was not tenable in Nigeria, given the fact that for many years Nigerians had not elected their governments. Regarding military rulers as pubic trustees would be a betrayal of the last hope of the people since military rulers had thrust themselves on the populace. This orientation to television service, though a feature in African societies due largely to the legacy of colonialism and the struggle for independence, is not democratic, as Ronning & Kupe (1999: 157 – 158) will argue. The other option, then, is that television operatives, as media professionals, would adopt the role of impartial custodians of the interests of a state, including its peoples. This means they should be independent of the political and economic elite, and be accountable to their professional ethics and ultimately the audience. To be democratic, the media should ensure that all significant interests are represented; that the public domain is not dominated or hijacked by particular groups. It should facilitate participation in the public sphere and contribute to the framing of public policy (Curran, 1993: 30). It has been argued that the media are organised according to the prevalent philosophical orientations in society (Siebert et al, 1976) but the importance of economic and political forces, along with the range of interests within the social context of operation, must be considered by Director Generals when interpreting their role and the relationship of their organisation with the government and the audience (See Curran & Seaton, 2003; Eldridge, Kitzinger & Williams, 1997 for accounts of the evolution of public service broadcasting in Britain). In government-owned stations, government is still regarded as the benefactor, so it is privileged.

Because Director Generals are responsible for policy decisions, they have to liaise with government. They also maintain links with the media and other cultural mediators, including self-appointed spokespersons for the audience. The Director General is the person who tries to balance the professional interests of the organisation, including those of the staff, with the interests of the government and influential individuals, some of whom may be self-serving *bounty hunters*, who are close to the corridors of power.

The Director General, plus the other Executive Directors and external parties, are members of the governing board, where such a board exists. The board serves as a bridge between the organisation and the community. It is chaired by an external member who is appointed by government. The board serves as a check on the Director General and helps to steer the organisation in appropriate directions. Members of the board are selected to represent diverse interests in society. Invariably there is an attempt to reflect the Federal Character and to ensure that minority communities have a voice. There is usually a voice to represent women and children. The board ensures that there is judicious use of resources and that policy is properly interpreted in accordance with the stated mission. The board is the de-facto supervisory organ for the authority, but the authority still has a peculiar relationship with the federal Ministry of Information.

Being a government-owned station makes NTA a parastatal under the Ministry of Information. Though this ministry has a supervisory role and can be quite influential, it does little in terms of direct regular interaction. The following description by a former director in the ministry, on the relationship between his ministry and the authority, reveals how the ministry can exert pressure on the authority if it wants. Control of the purse strings was the ministry's ultimate means of regulating the authority.

> The supervisory role of the ministry is one of acting as an intermediary between the various parastatals and government. If the authority wants anything from government, they put requests to the minister, who in turn presents them to government at the Council of Ministers. The request is put in [the] form of a memo to the Presidency through the council, and the response is passed back through the same channels to the ministry. Finally, [budgetary] allocations which are made by government . . . are passed through the minister – the ministry.
>
> Other functions of the ministry involve authorisation for overseas travels. Staff who must travel abroad to undertake

assignments, sports events, attend professional meetings, training courses, seminars and workshops must obtain written approval, through the authority, from the Honourable Minister.

That was the situation under a military government. The situation in the post-military democratic dispensation is little different. For instance, any expenditure that exceeds 20 million naira must be referred to the Ministry of Information for approval. If the expenditure exceeds 50 million naira, it must be referred to the Federal Executive Council. These are much higher thresholds of spending than the authority was allowed at inception of the organisation in 1977, and even in the early 1990s. The relaxed regime reflects an awareness of inflationary trends and the value of the naira, as well as an understanding of how constraining such regulatory mechanisms would be.

In the absence of a board of governors, as has sometimes been the case, the ministry is expected to "oversee" the running of the station, setting policy and granting approvals as the board would have done. The ministry is the final clearing house for certain matters. For instance, contracts of a determined value (as discussed earlier) even with an operational board in place, still have to be referred to the ministry. The ministry also supervises capital projects to ensure that these are completed to acceptable standards. The existence of the board should be a relief to the ministry, especially as regards matters of policy.

In spite of the scheduled protocol and the bureaucratic relationship between television and the ministry, it was not unusual for political office-holders to by-pass these and throw their weight around in the station. This could be done informally; sometimes a phone call is all it takes. As has been illustrated in previous chapters, failure to comply with the whims of the big-wigs could cost staff their positions, no matter how highly placed they are.

A range of administrative responsibilities fall directly under the office of the Director General. These include the office of the Secretary to the Board, the Legal Adviser and the Internal Audit Department. The Corporate Affairs department is also under the Director General's office. This shows that running a television business requires attention to policy and legal obligations; the financial wellbeing of the ventures; the assessment of projects and performance. It is also concerned with image-making and public relations, especially smoothing relations with the influential "big men".

Administration & Finance

Administration at the directorate level is responsible for staff recruitment, discipline and staff welfare. It ensures that corporate policy and guidelines are properly interpreted and implemented in the day-to-day operations. The directorate is responsible for appointments and promotions. Staff for graduate-entry level jobs and above are hired at the headquarters, then deployed to the various stations (production centres). The centres may identify and recommend individuals whom they would like to recruit but such recommendations must still be ratified by headquarters. Similarly, in matters of discipline involving senior staff, control has been moved out of the local context; any disciplinary action must be sanctioned at the directorate. Having centralised control over recruitment helps to maintain a balance in the Federal Character at the organisation. The federal character policy is to ensure that federal organisations are not dominated by a group of people from particular states; there are staffing quotas that must not be exceeded.

Administration (once known as Manpower Resources Directorate) used to be responsible for training, but the NTA created a Directorate of Training and Capacity Building in 2003. This development is a reflection of the magnitude of the burden of training. As an old generation of television broadcasters bow out and new technologies are introduced, training becomes a more serious imperative. The new vision for training is to "transform the NTA workforce into a world-class television network". Local and international training courses and industrial attachments are organised for staff. Employees have been sent to institutions such as the Administrative Staff College of Nigeria, the Centre for Management Development and the Nigerian Institute of Management, as well as foreign institutions, for example British, American and German broadcasters and manufacturers. (Other television organisations send their staff even further afield wherever there are organisations to collaborate with.) NTA Administration departments are still involved in organising courses where crucial work-related knowledge is shared. Training is facilitated by retired or current staff who have particular expertise that can benefit other members of the team. At such fora, best practices are identified and the institutional culture is established. The advantage of relying on retired staff is that they are able to reflect on past experiences while also sharing the vision of the institution. This could be restrictive in an industry as dynamic as global television is, if there is undue insistence on the "good old ways". On the other hand, it is most beneficial when there is adequate recognition of the limitations of old ways and emerging needs. Nigerian television operatives have not been reticent in adopting new trends.

The NTA has experimented with different management structures but sometimes it takes a while for staff to catch up with details of changes that have occurred. It is up to the administrators to maintain the smooth running of an organisation in times of transition, yet there is evidence that administrators sometimes find it hard to keep up. Zonal network centres, for example, were introduced to assist in speedy and participatory dissemination of information, especially from the grassroots. This objective is now being accomplished. However, the new system had some temporary administrative setbacks at the initial stages. Certain members of staff were ignorant of the impact of these changes on the reporting lines. A manager in Administration at one of the new zonal network centres was not certain how the new management structure affected staffing and assignment of duties.

> Before the new zonal structure, there were managers at the stations taking care of business in the different units. For example, the Manager, Marketing takes care of the station itself. There is also an Assistant Director, Marketing. I used to have a Zonal Marketing Manager who coordinated the affairs in the zone . . . now I do not know . . . I am not privy to what is happening at the headquarters. (Interview conducted July 2008)

Though there must have been memos and directives about such decisions from headquarters, if information does not percolate to the staff concerned, feelings of ambivalence may persist, causing apathy or confusion, even resentment. Much of this can be resolved with prompt, adequate and lasting communication. Written documentation that is within the appropriate public domain will promote transparency and reduce ambivalence. Officers in Administration, along with those in Corporate Affairs, are involved in documentation of developments. Commissioning of such publications often occurs at strategic points in the organisation's history at the instance of different interested parties. These publications highlight the affairs of the organisation, while the interviews and features offer staff a chance to hear directly from management. Access to such publications as historical record varies depending on the date of publication, the target audience and the number of copies in circulation.

The *NTA Corporate Newsletter* is a publication that addresses the authority's internal audiences and helps to disseminate news and information about the organisation and its staff to the workforce across the nation. It was designed to be a monthly information bulletin published by the Corporate Affairs department. Before this initiative there had been others, like *Contact,* which was described as the house organ of the NTA. *Television Journal* was another NTA publication but it had a different focus; it served as a professional journal, offering ideas from the academic community and seasoned practitioners. *TV Guide* was less intense in its orientation; it showcased the line-up of programmes from the various stations and highlights from some of these, as well as interviews with the stars of featured programmes. *TV Guide* was a means of communicating with audiences and advertisers. These are distinct from the *Corporate Newsletter* or *Contact,* which is a step beyond the noticeboard and internal memos. For an organisation of its size, one that is yet to make the most of information and communication technology, its website being underdeveloped, such publications are vital. Smaller television organisation such as Lagos Television use this technique to disseminate information, sharing corporate ideals with its stakeholders. The permanence of printed material makes it attractive for long-term documentation.

Fig 6.1: Montage of Nigerian TV publications

The Finance Directorate is responsible for the management of the corporation's financial issues, top of which are salaries and wages. Other areas of operations in its schedule of duties include monetary transactions, both internal and foreign exchange, purchases and credit control, sales and debtor controls. The directorate also covers management information and the accounting system. It is responsible for policies and plans that ensure probity, accountability and the judicious use of resources. Its policies affect the administration of the stations and the output of the organisation. With a large staff, the directorate has traditionally had to worry about a bloated wages bill. Yet, the core business, trading in airtime and ensuring that the right audiences can be delivered, formed a vital part of its remit.

In an interview with *Television Journal* (July/September 1985) Dr. Christopher O. Kolade, chairman of the federal government-appointed Committee on Rationalisation of the NTA, noted that his committee found a need to slash the NTA workforce to make it commensurate with its proper activities.

Kolade: That means a reduction from 9,700 to about 5,000.

Television Journal: Was funding the authority a part of the basis for rationalisation?

Kolade: We found that the government was talking to NTA about levels of annual subsidy in the region of 66 million naira, and already the previous year, the NTA personal emoluments was in excess of 60 million naira. Back then, station managers spent a fair proportion on staff welfare, including vitals like hospital bills for staff and their dependants. It is up to Administration and Finance to ensure that there is a right balance between such expenditure and that required for the core business.

Also crucial to the directorate are disbursement of funds and credit control. As with every business venture, stations have to maintain reasonable cash flow; the more complex the business is, the more challenging this is. It has been suggested that "almost all media businesses face five main challenges: continually developing new content offerings, addressing a triple market interface, coping with volatility, dealing with multiple local markets and balancing economic [objectives] with more social objectives" (Aris and Burghin 2009: 3). In their view, the triple market interface consists of authors, end users and advertisers. All these challenges have serious implications on the duties of the Finance Directorate (or department, as is the case with smaller stations). Regular outgoings from stations include the cost of local production. When stations are part of a network, they have to deal with multiple local markets, catering for audiences who also expect to be part of a regional national and global community. In such circumstances, the normal logic of cost control has to be considered against the wisdom of investing in less viable markets.

Content offerings for television stations, News and other programmes, are perishable. If the audience is to be satisfied, there must be adequate funds to ensure that production is not delayed or compromised, yet the audience is only one of the three interests that media businesses are concerned about. Audiences are the *raison d'être* of television business, the commodity to be exchanged for patronage of government or commercial patrons; therefore stations need to secure the attention of the audiences.

Of the vast range of creative personnel who make the content of television organisations, many are independents. Television organisations have to pay for the right to use the intellectual property of others. Ensuring that adequate funds are available to service these deals has become a greater imperative in the more commercialised industry that has evolved in Nigeria. Whereas it was once the case that the creative team made many sacrifices, there is now greater awareness of the value chain, the value of creative efforts, and the contribution of advertising/sponsorship revenue. Fiscal planning needs to ensure that artiste fees are decent, no matter how modest, and paid promptly. Independent producers and media companies that invest in productions (often without much support from commercial banks) expect similar considerations. Attracting the right sponsors for programmes has its own costs; this is part of the equation to be balanced by the Finance Directorate. To these are added the costs of maintaining structures and assets, including lands and buildings, furniture, fixtures, plant and equipment, recording tapes and discs and vehicles. These all form part of an organisation's fixed assets as shown in the DAAR Communications Financial Statements. (2008: 68). There are other operating and administrative costs as well as debts, which have to be carefully monitored and controlled to safeguard the liquidity and profitability of organisations. This is consistent with the fourfold function of a Finance Department outlined by Borisade writing about Gateway Television, formerly Ogun State Television (2006: 151). These include support for day-to-day operations, asset management, fiscal control, surveillance of market situations and organisational finance planning. This confirms that even organisations with a public service ethos must be well managed; their investments must be profitable. Of the range of responsibilities, debt control is perhaps the most challenging, irrespective of the type of clients the organisation attracts; whether they are government agencies or big business.

Directorate of Marketing

Marketing has acquired an elevated status in operations since the commercialisation of service in response to the dwindling financial input from the government. When the NTA network was the only effective means of reaching a national audience, the directorate was in a privileged position. The advent of state government-owned television stations and private stations has altered the market. Since the growth witnessed in the Nigerian industry, advertising funds, which are the traditional revenue base for television, are subject to more keen competition. Advertising business has expanded greatly since 1959 but agencies vary in capacity and competencies. Television is still a most effective medium for visual impact and demonstration to aid message and product recall. The market therefore remains promising. However, when clients have options—multiple channels to choose—from, stations have to work harder and be more innovative to secure their patronage. Television organisations, including the NTA, can no longer afford to be laid back about the sale of airtime. But for its network news transmissions, NTA's claim to exclusive large audience shares is being challenged. Marketing has to be more aggressive to seek new opportunities, woo new partners, seek new ways of generating revenue.

The Marketing Directorate is responsible for promoting the output of the organisation, whether from its News and Current Affairs or Programmes Directorates. For this reason, commercial officers were assigned to work within those directorates. Other activities of the directorate include the organisation of exhibitions, competitions, fairs and similar activities that generate funds and programming opportunities. It was even responsible for buying and selling rights to international events but this function has been taken over by another business concern established by the NTA - NTA Enterprises Limited.

Established in 1990, NTA Enterprises promotes initiatives that are allied to television service, such as the NTA lecture series, movie initiatives, music videos, Youth Leadership Awards, Queen Nigeria beauty contests and soccer shows, including the English Premier League and local football challenges. NTA Enterprises also promotes other types of entrepreneurship. It is involved in property management, car-hire services and management consultancies (for business centres and internet cafes). These are less traditional sources of revenue which television business now taps to maximise its potential. Other revenue potentials that are more directly relevant to the core business exist. It is now common practice for television organisations to offer different elements in the production chain to the public for a fee,. Stations will cover special events, grant access to their recording facilities, provide facilities for dubbing, design and help produce messages on other media platforms. All these show the revenue-generating potential of television business.

Fig 6.2-4: flyers of NTA promotional activities (2) The UEFA 2008 Cup, (3) Information Managers' Strategy Retreat and (4) 2008 Beauty Pageant

The NTA—which had been backed by the financial might of the federal authorities—seems to have taken a cue from initiatives evident in state government-owned stations, some of which can best be described as desperate. The most controversial of these is the designation of certain news items as Commercial News (LTP—Let Them Pay). Gateway Television (GTV, then Ogun Television, OGTV) was among those that pioneered this practice. They define commercialisation of news as the full reporting of events or pre-event mentioning of upcoming activities. Strictly speaking, some of these lack qualities that news editors would normally consider in selecting stories (Adebimpe, 2006: 240). These include social events, festivals and cultural activities, annual general meetings, and other activities from the business sector that would benefit from the publicity. At times stations were not bashful about charging for stories which might have made the news on their own merit, if the subject would benefit from the coverage.

This practice had been so controversial that even the National Broadcasting Commission (NBC) was averse to it on the grounds that the integrity of news would be undermined (Adebimpe, 2006). Another reasonable ground for objecting is the possibility that bona fide news stories are bound to be left out of bulletins in order to accommodate the more commercially viable stories, or if the subjects of items designated as commercial news stories were unable to raise the assigned levies. Yet the practice may have been an effective way of filtering out the throng of lobbyists who descended on newsrooms in an attempt to get coverage for their events. Consequently, rather than recede, the practice gained ground even at the NTA, which sought to uphold professional standards. News Directorates in conjunction with commercial officers ensure that commercial stories are so designated and they are not covered free. Indeed, it was partly due to the recognition of its revenue potentials that the duration of the main evening news, *Network News at Nine,* was extended from 30 to 45 minutes. In the 1990s, it was not uncommon to have an entire 15-minute segment of the news devoted to commercial news stories. The NTA has since extended the broadcast day to about 18- if not 24-hour service. There are more slots taken up by network news programmes in the broadcast day. These offer more potential for revenue.

As part of its efforts to recognise revenue opportunities, NTA's Marketing Directorate has been involved in commissioning research to appraise the organisation's performance. Though for good reason academics may view such efforts with scepticism (since the research findings were meant to assist in appropriate pricing of airtime) such exercises should not be dismissed lightly, as they become central to the management and operations of stations. The 1985 claim made by the authority that 30 million Nigerians watch the *NTA Network News at Nine* was based on one such research exercise. The study was conducted by the Research Bureau of Nigeria. Though commercially motivated, such research served as the basis for making policy and had an impact on output and other practices of the authority. The research was the first acknowledgement of a deeper penetration of television; a medium which had hitherto been regarded as elitist was assuming a more populist posture. The result was based on the logic that there were on the average five viewers per set for the 6 million sets available at the time. The sums appeared flawed in that the study may have overlooked several hindrances to television viewing, including the perennial challenge posed by poor power supply. The view was hotly contested in the media at the time, not least because it challenged conventional wisdom in the academic community that television was elitist. Twenty years on, and the medium continues to wax strong, with evidence of less conventional modes of facilitating reception, especially in the urban slums and rural towns. Little wonder that Marketing has taken a more central position in shaping the output of television, as the following case study will show.

Marketing @ NTA Kano

NTA Kano is a good example to consider regarding an interest in marketing as an aspect of television operations. The station is in the North West of Nigeria, further north than Kaduna, though it is designated within the Kaduna zone. Kano is an ancient city that remains a relevant market with its dense population and cultural (religious) sensibilities. As a local NTA station, it has access to various streams of funding, and marketing is crucial in tapping most of these. Sale of airtime must be first on the list. So how do the operations of a Marketing Directorate or the local Marketing Department impact on the activities of a local NTA station, especially one that is far from the centre of the large organisation?

Airtime is typically sold to advertising agencies but because they usually have accounts from the big clients that manufacture goods and services and seek national audiences, their bookings are done centrally through the Marketing Directorate. The media placement orders indicate the number of spots and the required pattern of placements which agencies want. In other words, the order indicates the local stations required. Placements are done on a credit basis. Each station raises invoices for the number of spots that it has run. Payment is expected within 60 days.

Perhaps it is for its troubles in administering the transactions, sourcing the ads or because of its responsibilities to the other less viable stations within its network, but the central authority is entitled to 60 per cent of the funds, while the stations should receive 40 per cent. (Station managers mutter that they may not even get this proportion of the proceeds of sale of their airtime.) Sale of airtime from network slots is thus not adequate for the running of a station.

Stations may have got by in the days before commercialisation, when they were virtually fully funded by government, but not in the era of commercialisation when each station is meant to generate the bulk of its operating costs. That stations still rely on subventions from the headquarters (for capital expenditure and staff remuneration) shows that local revenue generation is inadequate. New expenditures have emerged at the local stations since commercialisation. Central among this is the issue of power generation. Stations can hardly afford to buy fuel for the electricity generators used to power stations, even with all the strands of revenue available. The revenue from commercial activities is a real life-line for a number of stations. This makes the proportion of revenue accruing to the headquarters appear excessive.

NTA Kano is on air for about 14 hours a day. It cannot afford to run for 24 hours daily, as recommended. During transmission hours, it runs on a generator, burning diesel that costs up to 200 naira per litre. It needs the revenue from the sale of airtime but its biggest clients are the advertising agencies, and government, which operate on credit. Placement orders from these sources are often processed through headquarters. Though patronage is high, given the number of messages placed by these clients, what is due to the station is a mere fraction of what the organisation receives, when it eventually receives it.

Paid announcements and advertisements from local sources offer the station greater control. It has to work harder to generate these, as local clients are of the view that television is so expensive, it is beyond their means. But managers at the local level have been flexible. They apply discretionary rates; negotiating deals that would suit local advertisers and benefit the station. It takes a manager who knows his market and the business to recognise opportunities where discretionary rates can be applied to the advantage of the station. The 15 per cent agency commission which advertising agencies enjoy but which should not accrue to independent clients is an example. Such consideration encourages local businesses to bring in bulk orders like advertising agencies, if they can be convinced of the benefits of advertising. The station even hires agents who canvass for adverts in the community to ensure that local businesses are aware of these benefits. These agents scout for patronage from among those who would normally not consider that their wares are worth promoting on air—local entrepreneurs, market traders. To promote sales and ensure that the station survives, management negotiates deals; trying out novel schemes, offering discounts. Managers even offer "credit facilities", just to stimulate the market. The experience in NTA Kano and elsewhere suggests that the formula works.

Programme sponsorship is another crucial source of revenue. Programmes (such as drama, musicals, quiz shows) and advertisements or other sponsored messages (like public service announcements) often come pre-packaged, syndicated in the selected markets. The increased activity within civil society has increased the volume of business that stations can thus receive. Non-governmental organisations, community-based organisations, religious bodies and such institutions with a cause to promote tend to make their own programmes and distribute it in this way, paying for airtime to the stations. Stations have the responsibility of previewing the material to be broadcast. Programmes must meet the technical and regulatory standards for broadcasting if they are to be acceptable; if not, no sale. Commercialisation still has to be within defined parameters. The cultural/religious guardians in Kano are among those who raise the bar higher than what is the average for the nation.

Programme sponsorship has also thrived because of the activities of independent producers, who tend to focus their attention on cutting deals with the Lagos-based big spenders. Some of the programmes made for national audiences, especially those supported by trans-national agencies or intergovernmental organisations, tend to have the resources to carefully plan and produce messages that will be sympathetic to a wider range of cultural sensibilities. Often these are drama productions, which means that stations have less incentive (in terms of time and resources) to produce dramas locally. In any case, drama has become even more expensive to produce, as the experience in contemporary Nigeria shows. Artistes have become more commercially motivated since the boom of the domestic video/film industry, Nollywood. Kano has its own specific niche of this industry which some refer to as Kannywood. Prior to this development, artistes were willing to perform just for the pleasure of being involved. They would be grateful to receive a token allowance to cover their transport costs. Not so these days. Artistes are conscious of industry rates, so people request fees of up to 60,000 to 70,000 naira per production. Acting, for many, is no longer a hobby — many have turned professional. Some artistes have even secured university degrees to set them on their career path. The cost of productions has risen as expectations have. So when the stations produce dramas locally against all odds, they must struggle to find sponsorship. The support needed may not be locally available, but all is not lost.

NTA Kano, like many other commercially minded stations in the south, has identified another stream of funding – facilities rental. This is big business in state-government owned stations such as Lagos Television and Ogun State Television. These stations have multipurpose halls that they hire out for, wedding receptions, birthday parties and meetings, which are featured in designated programmes for a fee.

Studio facilities are another category of facility for hire. NTA Kano makes these available to sponsors who want to produce programmes locally in order to make them more relevant to audiences in northern Nigeria. Such producers have been attracted to use Kano as their production base because of its leading position as the largest commercial city in the northern region and the reputation of its markets. An example of such a production is the audience-participation programme *Taishi mataimakeka* (Wake up, let's help you) which is recorded in the NTA Kano studios. This Hausa-language programme has been compared to *Who Wants to be a Millionaire*, which is available in English on other stations. Prizes on *Taishi mataimakeka* can be life-changing for the audience it serves. People have won as much as 100, 000 naira in cash, or motorcycles. The questions are on local current affairs. In this way Marketing ensures that the facilities are put to good use, programmes are generated, audiences are well served, and the station is endeared to its market.

NTA Kano takes pride in its studio facilities. While the brick and mortar erected at its inception in 1980 remains solid, the equipment had run down and become outdated. For it to remain an income-generating asset, it was necessary to plough back some of the earnings from the commercial activities to upgrade the technology. By 2008 the station had gone digital, using non-linear editing and mini-DV tapes; gone are the days of celluloid film and reel-to-reel technology. No more U-matic and VHS tapes. Dembo, the General Manager at Kano says these accomplishments could not have been possible had he relied strictly on the subsidy from government or centrally generated funds.

To think that Marketing was once the dead end unit of television business! That was back in the days of heavy reliance on government subsidy. Experience has shown that in the more aggressive climate that has swept into television broadcasting in Nigeria, Marketing must be regarded as the powerhouse; it will ensure that the other engine rooms can keep working; that there is a ray of hope for the tube.

Figure 6.5: Multi purpose hall for private hire

Directorate of Engineering

The Engineering Directorate is responsible for the formulation of the NTA's corporate policy on engineering. This includes giving recommendations on the standard of equipment to be acquired and the standards to be maintained in transmission. To do this effectively, it must keep up with professional developments within the engineering industry. Since much of this engineering activity happen abroad, the Engineering Directorate liaises with external suppliers and professional bodies.

Its staff may be least visible to the public, but it is the most indispensable arm of station operations. To ensure that its service is effective, the Engineering Directorate provides guidance for engineers in the field, in stations across the nation. These are headed by chief engineers (or managers).

The directorate is responsible for planning, designing, research development, project initiation, implementation and operation. The configuration of transmitters to ensure effective coverage in the designated market areas is a core aspect of its business. It is responsible for all the transmitters, their maintenance and the procurement of spare parts. Its activities are geared towards ensuring the acceptable technical quality of the stations. It is responsible for the maintenance of studio facilities, vehicles and other operational equipment such as outside broadcast (OB) vans.

Many NTA centres can boast of having standard studios, but this has to be more than physical space. Standard studios have to be defined by the available equipment, not just the bricks and mortar. This is important, as many of the Nigerian television studios, especially those built in the first wave in the 1960s and 1970s need to keep abreast of the latest technology. But keeping up to date with technology in such a field as television is difficult. According to Jimmy Atte, who at different times served as Executive Director both in Programmes and News Directorates,

> The technology killed us. The changes in technology are so rapid that it is difficult to keep up, yet the pressure to do so was from several sources. Some are from within and some are external. Take the recording equipment. To begin with, there was the telecine. All news reports were cut on telecine in those days. Then there was the two-inch tape, and afterwards came the U-matic (¾-inch tapes). The two-inch tapes were regarded as having better quality. It was assumed that a change to the ¾-inch format would lead to a drop in quality. When the sources of such views could be traced to the BBC, an institution which was well-respected and one from which some had received their training, this view was deemed to be canonical; but there was more to that assessment than is readily apparent. It turned out that position had been informed by the peculiar institutional experiences of the BBC at the time. The quality was great but this technology was quite restrictive; it was either studio-based, or you required outside broadcast facilities. Because [of the camera chain that was required for this sort of tape] you needed more people in the [production] crew. The format was cumbersome for production. At that time you required a bus or two station-wagon vehicles to convey the crew to the production site. A change from the format would have had serious implications for staffing in the organisation. [Interview conducted in 2008]

Atte went on to explain the adoption of U-matic cameras initially posed a risk to some staff positions. Even though they were bulky in the early days (compared to digital camera chains available today) the

U-matics were more flexible. That you could get by with a crew of three people was a serious advantage of the technology. Yet experience had shown that two-inch tapes were far more durable. They were more rugged and may have been more suited to operations in Nigeria. However, once the global industry had shifted gear, all had to change in Nigeria. Television stations in Nigeria have no clout to influence the wider industry and are always having to catch up.

Because the nation relies on imported technology, engineering spending tends to be a huge drain on resources. Yet without the requisite investment in appropriate equipment or spares, television stations cannot thrive. Lack of spares and obsolete technology ranked high in the complaints from stations during field studies in 1991 and 2008. In the face of dwindling foreign reserves, this technological dependence is a dire problem that impinges on day-to-day operations. The following excerpt from an interview with an Engineering Officer illustrates the point.

Q: Is it really necessary that we must have the latest in technology?

A: Well, not really necessary to have the latest in the market, but we must have the ones that are still being produced, ones that we will have spare parts for.

Q: So it's a matter of spare parts?

A: Yes! It's a matter of spare parts because the average life span of most equipment is about 10 years. When it has lasted about 10 years, then the manufacturer will [pack it in and] say, "It is under." It has become standard practice that they will keep on producing that spare part for the next 10 years. So they are not committed to any consumer to produce the spare parts [thereafter]. If you have to go back to procure the spare part, you have to pay extra for it. It's got to be custom-made for you and you've got to pay more for it. So that is why we must really keep up to date.

Television is thus an expensive venture, whether the nation attempts to keep up with global trends in technology or it opts to cut corners by going for fairly used equipment that may soon be discontinued.

There is, therefore, a greater need to look inward. Local ingenuity in solving engineering problems is desirable but this is only if it will not compromise universally defined technical standards. In reality, this may be far from happening, as there is inadequate financial investment or encouragement for such developments from the government or private sector.

Although the engineers have been privileged to produce Director-Generals of the authority, there is yet a lot to be done on the image of television engineering. Engineering and technical services form the backbone of television service and are easily blamed for shortfalls in the service. Over the years, the audience became familiar with apologies for breaks in transmission "due to technical faults". Such claims were sometimes justified but they were often the excuse to cover lapses in other areas of operation. Blaming technical hitches was an easy default for duty continuity announcers. As the technical arm of the service, engineering is overshadowed by the glitz and glamour that accompany the other specialisations. The situation is not peculiar to the NTA. It is a challenge for Nigeria's television industry. Indeed, it is a reflection of the state of engineering in the wider Nigerian society.

The following account is an engineer's description of his department. Although he spoke with regard to his experience in a state-government station, the account illustrates the range of duties performed by an Engineering Department in a station; its responsibilities to other departments and the (dis)ingenious means this engineer resorted to in order to meet these. The statement is a description of what happens at a morning meeting following complaints of weak signals from the station; a situation that was prevalent in the early 1990s.

> . . . Let me tell you in a nutshell. The Commercial [Marketing] man will be fighting [at odds with the engineers] because people are complaining that they are not receiving us [signals from the station] . . . somebody in News, a newsman will be fighting you for a vehicle – that you didn't give him a vehicle for a particular time. Now somebody in Programmes will say, 'You have given me a bad camera. I've gone to location for about three days; come and see the [poor]

outcome of the production'. It's like a madhouse; everybody's at the throat of engineering.
[Interview conducted in 1991]

In almost two decades since the above observation was made, not much has changed. Therefore it is worth pausing to note the problems faced by such an integral and self-effacing arm of the business.

The Problems with Television Engineering

The following is a compendium of some key comments relating to engineering that were raised during the observational study of the stations. Much of this has to do with training.

Television operatives often complain of an inadequacy in the number of trained technical personnel available in television. Although several institutions and colleges offer electronics or allied engineering courses, their graduates still need to be trained on the job to help them adjust to the peculiarities of the industry. Without access to equipment with current industry specifications, fresh graduates may have no more than principles to apply. This is a good place to start but they must build on this, hence the need for appropriate exposure and training.

When new equipment is installed, staff are expected to be trained by the manufacturers. This may entail overseas study trips for some core staff, who on return will pass their knowledge on to others. As an alternative, the manufacturer's representatives could be flown in to offer induction courses on the equipment. The complexity of the kit may determine which option is more cost effective. Yet in the effort to reduce costs, or in the event that middlemen choose to maximise their gains, arrangements for training may be compromised.

The NTA Staff College offers training and courses that help to orient and develop engineering staff, but unless staff can be released from their duties, they may not benefit from these. For that reason, the tendency is to rely on in-house training. This way, traditions and nuanced ways of operation are more likely to be perpetuated and ingenuity may be stifled. Yet all these are a fraction of the real challenge, which is that the television industry is unable to offer adequate incentive to attract a large corps of qualified engineers.

Remuneration in the television business does not in any way compare with the salaries offered in sectors such as oil and gas, banking, finance and even mainstream industrial sectors to which engineers flock. With their drive and ambition, young engineers who start their careers in the TV industry are hardly enticed to stay on. One engineer described some of the experiences that drove him to quit.

> They operate like ministries [ie, civil service]. They operate the same salary, you know salary scale. You can't really keep professionals with that poor salary.

Q: Was that why you left?

A: That wasn't the [only] reason why I left but one of it (laughs). [Interview conducted in 1991]

He further explained that he went in search of greener pastures; more experience, more exposure and greater challenges.

> There's a colleague of mine, he's still there. I met him a year ago or so. He said, 'Ah, I'm still at NTA, oh! but I'm still trying to find a way out. I want to go out'. Although he has risen very well, if he's not a Chief Engineer now, he'll be a Deputy Chief, but he's still looking forward to leaving because he's not getting that challenge too. He said, 'Well, I just want to see whether I [can] complete 10 years [ie, qualify for his long-service award], then I will go'.
> [Interview conducted in 1991]

Though the issue of pay has been addressed, especially during the Obasanjo administration, it was still a sore point with many workers. It was one of the reasons given for the strike action by the Radio Television Theatre Workers Union (RATTAWU) in July 2008. Although this was not restricted to staff in Engineering, it is indicative of their concern, as described by a union leader (who, incidentally, is also an engineering worker).

> The current strike action is evidence of changes in staff welfare. Staff welfare seems better but the inflationary trend undermines

improvements in the pay. The strike action is an expression of grievances over the two-year delay in the payment of the monetisation arrears promised by the establishment . . . NTA is doing fine but we do not like the Civil Service structure. It is one of the issues. The NTA apparently was on the same salary structure as the Central Bank but after Udoji [the 1972 workers' salary review] NTA was *pulled down* to join the civil servants.

RATTAWU does not seek to antagonise government, it will only seek to defend the welfare of its members – salary structure, prompt payment of salaries, [provision of] loan facilities."

[Interview conducted in July 2008 – my emphasis]

The above excerpt requires some background information for a better appreciation of the changes in employers' perception of the television worker. When government alone was involved in the business of television, television workers were public servants. Initially they were on a separate scale, like the staff of Nigerian Telecommunications, Central Bank of Nigeria and such essential services.

Prior to the harmonisation of salary scales, the nation had different labour unions, each making a case for special dispensation in view of the particular nature of their job. A harmonisation of the salary scales was meant to put a stop to those agitations. The NTA was brought under the Civil Service structure, a development which is detested by many, especially with biting inflation and the rationalisation of jobs. If RATTAWU had its way, NTA staff would return to a more privileged scale, such as that enjoyed by Central Bank of Nigeria staff in recognition of the inconveniences that go with their jobs. Submissions had already been made to a government committee on salary structure—the Shonekan committee 2006—to this effect.

Clearly, in contemporary times, an engineering job is not lucrative in this context;,and it can be rather frustrating. Even when there is money to procure equipment and spare parts (and there is never enough), the bureaucracy involved in the operations, and the delays that these constitute for planned maintenance make the culture at work undesirable. As mentioned elsewhere, when the costs of procurements exceed a certain value, approval from the supervising ministry is required. Also frustrating is the need to chase manufacturers for parts for discontinued models which stations has been saddled with because that was all they could afford. Sometimes this is against the better judgement of the professionals who see the wisdom in a smarter pattern of investment but have to yield to the competing demands for funds. Added to this is the waiting period that may accompany most shipments and clearance of haulage.

What is not included in the official story is the human factor—the culpability of individuals within the chain. Can the system be immune to the less savoury practices that are rampant in the larger society when it comes to tenders and contracts? This is mere speculation, yet if corruption has pervaded the land, it may manifest itself in decisions and how priorities are set. If quotations for contracts are inflated, so that substandard materials are deliberately procured; if gargantuan projects are pursued so that some may benefit, whereas this was not the area of greatest need; if available resources are not wisely disbursed in the greater interest of all, then the effectiveness of television services will invariably be hampered. Problems on the job are exacerbated—and they often occur sooner than later.

Another dimension of this human factor is the collusion among some ill-motivated officials who cash in on the situation in the public service sector to enrich themselves. Stories have been told of equipment, particularly vehicles, which are deliberately written off and auctioned to interested officials when they are still in fact serviceable. Some of these are serviced and put to use by their new owners, to the knowledge of everyone. Such officials had been known to sabotage the ingenuity of others who might have tried to salvage the equipment for their organisations rather than writing them off. This working culture is not unusual in the public sector, where individuals who seek to protect government property may be alienated by colleagues. Some glory in this activity, which is summed up in the saying "Government property is not my father's property."

The impact of such sharp practices is better appreciated from the perspective of the user departments. What would happen if an entire directorate or department is left with two functional vehicles while other vehicles in its pool are grounded? This has been known to happen, so producers are left to share vehicles, use their own, rely on public transport or be grounded. Any of these options would add to the logistics and cost of production. It is even worse if the News Directorate is left with only two functional cameras to produce a 45-minute newscast. This too has happened in the past, and though by 2008 there was euphoria about new investment in technical facilities, unless there is a change in the maintenance culture, the joy may be short-lived. Ultimately, lapses in engineering show up in the output of television stations.

Engineers also feel the pinch of poor decisions, and some have been known to take unusual steps to avert the situation. At one of the state-government stations, an engineer confessed to acts that may be described as sabotage. He took unilateral decisions to vary the wattage of transmission to extend the life of some elements within the transmission chain.

Q: Why don't you stay alive always [ie, why is the station not always on air]?

A: It's not economical for us. It's not wise, because it's like when you have your hotplate on at a low temperature, you know the element will not glow, you see. [The brighter the glow of] the element, the faster it burns out. So [it is] with the transmitter . . . At times we know we might run into problems, you see, we have peak periods . . . peak periods in the sense that when we have commercials rushing in, [but] they tend to fluctuate . . . you know, during our operational meetings, somebody from Commercial Department will be saying, 'My sales are low and it could be due to [poor reception]'. The newsperson will say, 'I went for an assignment yesterday. This is what people are saying. . . [complaints about poor reception]'. But as an engineer, you know why. You may not be able to own up that 'I lowered the wattage' because they might not understand what you are doing. But occasionally I do it. At times I come as low as 20 kilowatt if I know I do not have a spare part to hand and I don't want to flog a particular unit, so I just come as low as this. So somebody at Akure may not be able to see [the station] for that period. It might be for one week. So before he can come over [to the station to complain], we are back on air . . . He has an option anyway . . . He might tune to . . . [stations closer to him], maybe he views [this station because he] enjoys the drama . . . You [have to] use your discretion because most of our equipment [is] ordered . . .

from abroad and it takes weeks, and then [it is purchased] in hard currency too. So, unknown to them, I might be laughing at . . . [Commercial Officer], for instance, because I know I am trying to save some money [that is more than he is trying to gain], He might be thinking of 70,000 naira income. Okay. I will be thinking of saving about 300,000 naira.

Q: So at the end of the day it's economics?

A: Yes, it is. It depends on your handling of that equipment to know that.
[Interview conducted in 1991]

Many professionals may find it difficult to survive in this kind of system. Those who are attracted and remain may do so only grudgingly, but times are changing.

By 2008 there had been major investment in production equipment and transmission facilities at all the stations studied. This is a relief for engineering staff and others. Such investment has facilitated the big strides in local productions as well as network and international transmission where applicable, but the old challenges persist. Even in 2007, according to the Executive Director of Engineering in an interview with the NTA Corporate Newsletter, transmitters had been purchased without relevant spares because of limited funds. The contract to train staff in the maintenance of the transmitters had not been an integral part of the initial deal. Subsequently, when the transmitters were "grounded", stations were confounded. But what is befuddling is why transmitters from such a reputable company as Rohde & Schwartz should be so problematic. The website for this German-based company shows that in its 75 years of operations, its business has expanded worldwide. The UK, Canada, Belgium, South Africa, even Egypt have local contacts for service support. The company website details a variety of sales and service arrangements that should keep waiting times to the minimum. It even offers live (24/7) support. However, there are no local sales and service centres for Nigeria, as is the case with most of Africa and South East Asia. The contact office to which they are attached is in Germany. This is not so convenient—but the bigger question is why the transmitters seem prone to faults.

Three reasons can be deduced from NTA's Executive Director for Engineering for the unreliable transmitters; poor operational practices, poorly trained operators; poor public power supply.(Amana in interview with Okereke & Rwang 2007). He explained that some operators fail to adhere strictly to the manufacturer's instructions. He observed that some transmitters had been running at full power for up to four years. In the event that the operators lack the parts, equipment, facilities and the training to maintain them properly, those transmitters may be flogged to death. This is often the case; especially when compounded by the spikes accompanying public-power supply. The transmitters are merely helped along the path to an early grave.

Contemporary practice in Nigerian homes and industries is that electrical appliances are hooked on to uninterrupted power supply (UPS) devices. In some cases (as in AIT) the UPS for the transmitters is connected to another UPS to safeguard the life of the transmitter and protect the investment. Most NTA and other stations transmit only on power from their generators. Public power supply is merely used as a back-up for non-essential operations. This development reflects the breakdown in public power supply. Though this has not always been the case in the history of television, this deterioration in public infrastructure alone has added to the costs of running a station. Investing in new technology was prompted by the digitalisation of broadcasting services in readiness for the 2012 deadline for digital transmission. This is another example of Nigeria having to keep up with global trends in engineering.

The competitive climate that has dawned in the industry in the third wave of television in Nigeria is another incentive, giving organisations the push to invest in engineering. There now appears to be an increased sensitivity to the fact that engineering holds the frame within which other departments must fit. The following is the testimony of an engineering worker from NTA Kaduna. It typifies the optimism that was echoed regarding engineering services across stations around the nation. Appraising television's role he said,

> This is basically an engineering perspective. In television the most important thing is to entertain, inform and educate . . . In live coverages for the past 15 years, we were totally handicapped; totally handicapped in the sense that, compared to now, most of the

information didn't reach the rural areas. Now, because of the creation of more stations, zones, zoning and . . . procurement of OB vans .
. . NTA can go to any area and transmit live there with the help of the Digital Satellite News Gatherer (DSNG) and the big [OB] truck. Before, we did not have such equipment, so things are far much better now. Far, far better!"

"The network centres Kaduna Abuja, Ibadan Jos Maiduguri [as well as Benin, Sokoto, Enugu] in the zones can now anchor the news, whereas Lagos used to be the origin of the news. This is another development. There is evidence that the audience appreciate the visibility of these other areas, from the letters they write and other feedback which comes during interaction with them in the course of programme production. Even former DGs confirm that we are privileged to have improved technology to work with.

"As engineering staff, I can confirm that some of the drudgery in the work has been eased with the acquisition of modern technology. For instance, during live transmission, you can go into your [DSNG] van and switch on the equipment, make your line of sight, get Abuja signal, confirm if Abuja is receiving your signals, no problem. But before, when we relied on links equipment, we had to climb the mast, we may have to pound the gear into place. It could take us up to one hour 30 minutes just to get the line of sight, but now it is not like that. It is now just as easy as possible. The digital satellite technology is far better.

This affirmation from an NTA engineer suggests that there is good cause to have higher hopes of a brighter future.

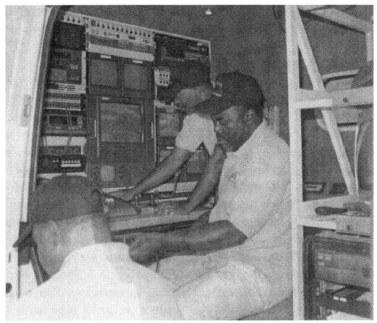

Fig 6.6: Inside DSNG van preparing for first transmission of Network news from
NTA Zaria
Image Courtesy of NTA Zaria Management

The News Directorate

Unlike the other directorates, which were initially conceived to be restricted to policy-making and monitoring units, the News Directorate was intended to be operational. It had fully fledged news operations at headquarters in Lagos (and later in Abuja). This was in addition to the news operations in the local stations which serve as bureaus, feeding into the central productions at headquarters as appropriate. This arrangement shows the importance of news to the organisation in the fulfilment of its mission; news was central to the effort of forging unity in the nation. This section will, therefore, focus broadly on news production practices. It will be concerned with the discrepancy between the claims of the organisation and its reputation; what accounts for the distinction between poor opinions held among sections of the audience regarding NTA news, and the organisation's pride in its news being balanced, credible, constructive and consistent.

In these claims is a hint of the operating philosophy; an explanation of the organisation's sense of duty, which many Nigerians will contest.

NTA Network News Objectives

The NTA (like other media in Nigeria) is statutorily required to "uphold the responsibility and accountability of the Government to the people of Nigeria" (Section 22, Constitution of the Federal Republic of Nigeria, 1999). All broadcast stations have a duty to report, analyse and comment on developments and issues of public interest. The Nigerian Broadcasting Code requires stations to act responsibly in their packaging and presentation of the news; News and Current Affairs programmes are expected to facilitate a clear understanding of issues, and an appreciation of various perspectives to these must be offered without bias and prejudice (Nigerian Broadcasting Code, 2002: 60 – 62). These norms are rooted in the legacy of the pioneering stations which had adopted Western ethics. According to Sotumbi (quoting Maduka), the expatriate managers at the premier station, WNTV, "were concerned with ethics, professionalism, commitment to excellence and editorial independence" (Sotumbi, 1996:173). He explained further that the adopted principles and values also reflected the Western Nigeria Government Broadcasting Law of 1959. This required the station to be impartial in respect of political or industrial controversy, or public policy. The NTA Code mirrored these values.

Network news service should be judged by the objectives set out for it. The legal instrument establishing the NTA states that its success will be determined by how well it delivers on the following:

a.   What the people ought to know. What the editor decides (guided by the rules).

b.   What the people anxiously look forward to, a daily communion, rewarding, refreshing, gratifying the experience of the viewer.

c.   Exploratory, in-depth (if not breadth) a face-to-face realism.

d.   A confirmer of earlier stories, or a dispeller of erroneous stories. the authority, credible, trusted and balanced.

e.   News for Action, for positive desirable social change in society – champion causes; such as the long-term strategic one of unity; anti-apartheid stance self-reliance; the Black man's burden, etc. (NTA Code 1981: 58)

It goes on to acknowledge the need to promote national unity, suggesting that network news adopt a slant towards national ideals. Yet the news is expected to avoid outright partisanship, even though the code considers a complete absence of bias as being naïve. This is understandable, given that all its operations are open to interpretation, beginning with the selection of the subject matter. The news is to cover a wide spectrum of subject matters,—whatever affects the viewers, and stories about different tiers of government. Still, as much as possible, the NTA seeks to be a mere conduit of information. Try hard as they may, though, NTA staff are aware that their service is often regarded as a forum for government apologists (Iredia, 2004:10; Sotumbi, 1996:177). The NTA is deemed to be His Master's Voice. Incidentally, the same is true of other government-owned stations; even private stations may not be exempted from this charge. There is evidence in literature to suggest that this tendency is not restricted to Nigeria; news organisations are generally susceptible to a range of pressures.

> Powerful political elites may pull formal levers or apply informal pressures to promote or restrict the circulation of information through news media; politically marginalised groups may try, perhaps in vain, to use information as a bargaining chip in a complex series of exchanges with journalists; news beat correspondents will seek to sort the valuable 'contextual information' upon which they may lay claim to a specialist knowledge, from that which they regard as contaminated by political spin. In each of these examples information is deployed through a variety of social practices and more or less consciously devised strategies.
> (Manning, 2001: 19)

Nigerian media may have their peculiar issues, but, as the above shows, so do other news media around the world. It is only through an understanding of processes in these organisations (in this case the News Directorate and newsrooms) that one begins to appreciate the value of the service. That the NTA considers the safeguarding of national unity as its primary duty, regardless of its other responsibilities, explains its practice. As a public broadcaster, the authority is concerned that its activities do not undermine social cohesion and national security. Though it is preoccupied with reporting government activities, it is of the view that by so doing, it focuses on the people's story. Similarly, it believes that by explaining government position, it offers information that people need to know. After all, informed people can better appraise the decisions and actions of their government. If this is the case, the broadcaster does contribute somewhat to making the government accountable.

The fact that a station is an official government organ makes it dubious, no matter how good its intentions are. Evidence over the years, including the summary dismissal of very senior officers, shows how susceptible stations are to external pressure—from government officials; the supervising ministry; Ministers of Information; politicians and other members of the elite, as discussed elsewhere in this volume —and helps to justify this view. But the privileged position of these stations makes them privy to a lot of information. The handling of such information is a grave responsibility, hence the authority's attempt to be constructive. In this attempt, how much of the privileged information it discloses is evidence of the authority's perspective on the public's right to know. This goes to show that access to information alone is not enough to guarantee credible and balanced reports. An organisation's perspective and treatment of the issues is therefore a more reliable clue to the quality of its news service.

There is evidence that the orientation of the organisation has improved since the democratic dispensation introduced in 1998. If nothing else, the *Manual on Political Broadcasts* (NTA, 2006) has been published. This sets parameters of operation in a range of situations. It lays out the minimum professional standards required of television practitioners. This document outlines the different personnel to be involved, their responsibilities, and it stresses the self-regulatory nature of the broadcast practice. In this is evidence of a determination to break from its poor reputation. In spite of such efforts, there are people even within the NTA who are critical of the way the organisation interprets its role.

We are still dependent on coverage of events. Agreed, this is what makes the news, at least for now, but **we have lost touch with human stories, investigative journalism,** follow-up, we are just a conduit for the government activities, relying on press conferences and invitation events. **There should be more spontaneity in this business,** but the administration is stifling the business. [Interview conducted in 1991- my emphasis]

News should be more than what government functionaries want. [By concentrating on this alone] **we are losing sight of the contributions of individuals who contribute to human development.** We are not accentuating the positive contributions and showing how they make a difference. **We should be able to follow up and ask why.** [Interview conducted July 2008 - my emphasis]

The similarity in the views expressed inspite of the time lag speaks volumes, yet there is a little comfort in noting that people-oriented stories are not completely avoided. The Sunday evening news programme *Newsline* is noted for these people-oriented stories. The News Directorate has tried to reduce the number and length of government-related stories, but the emphasis in its initial premise of operation is still betrayed.

While trying to appreciate the influences that frame the construction of news, the action of journalists should also be considered, as a General Manager of an NTA station, who had been a newsman himself, said.

. . . Let me be very frank with you. At times we create problems for ourselves. No government will ever come out to say: 'You must do this'. We tend to self-censor for the fear of the unknown. Basically, a reporter is a reporter all over the world. If you are a professional, you are trained, and you [are to] operate within the ambits of the professional ethics . . .
(interview conducted in July 2008)

The overwhelming evidence shows that reporters do have dilemmas on the job. In the course of their official duties, they may be constrained to go against their consciences. Like many other citizens, they are aware of the injustices in society, even when this occurs within institutions that ought to be challenged and made to be accountable. Though the rule of law appears to be flourishing in Nigeria, there are still a number of dubious court judgements. By virtue of their position in society, journalists are privy to the political manoeuvrings that attend such decisions, yet they are unable to challenge these openly. In the event of a blatant miscarriage of justice, journalists have no choice but to report what the courts have said. Though they could balance this out with the views of those who oppose the judgement, carrying news that is detrimental to important political personalities may prove to be too costly for the organisation or even senior reporters. However, if the personality involved has fallen out of favour in the appropriate circles, such a personality becomes fair game. One such case had occurred just about the time of the field research in 2008. It was regarding an appeal against electoral elections. The following is a comment made by a Manager of News at an NTA station on the matter.

> The moment you air the story that people are not happy with [the] judgement in favour of [member of government—name withheld], I bet you I will not last one hour here.

Television journalists are aware that theirs is not a peculiar experience; this happens widely in journalism, even the privately owned media. It is assumed that all journalists dance to the tune of proprietors and their cronies. The same Manager of News confessed his role in perpetuating this culture.

> If I own a medium, you dare not go against my editorial policy. I will use it to witch-hunt my enemies, to hound them, especially political enemies . . . Every medium has its purpose in this country. There are some that are influenced by their regional affiliations. There are papers that are pro-North, there are those that are pro-South and will never see anything good in the North. And those of us who are owned by the federal government . . . we just have to continue '*megaphoning*' what will suit them. So that's the way it is. That is the real issue . . . the way we operate.
>
> At times you file in the story and you go to the club, a public place. You are sitting down . . . people don't know you, but you can hear them saying [when discussing the news]: 'Look at them. NTA, stupid people!' You feel like hiding your face. It's sad. There has to be a reorientation, and this is more than training of staff."
> (Interview conducted July 2008)

In this manager's view, the situation calls for an appraisal of the pattern of funding. He (and others like him) suggested that ownership control has to be severed for journalistic ideals to be realised. Without a balance of control, the finely carved-out editorial policies are mere documents that do not reflect the actual practice. But there are other routine processes that reveal imbalance in the structures of news production.

The News Directorate is responsible for all the News and Current Affairs programming transmitted nationally. The linguistic complexity of national audiences marks out the orientation of the service. All the programmes from the News Directorate are produced in English. News on Nigerian television has thus been viewed as elitist, based on the boundaries of those who benefit from its operations.

On local NTA stations, newscasts are produced for audiences at the grassroots, to ensure a balance in service. Local audiences should not be overwhelmed by national and global issues alone. In order to reach these audiences, highlights of the national stories are translated into local languages; but the stations do much more than this. By taking on the burden of producing national and global news, the directorate places the task of producing local news in the hands of the local stations. This should be an advantage in the competition that now exists between local NTA stations, state government-owned stations and private stations to cover the localities. It ensures that there is a plurality of perspectives being offered, even if the issues they cover are more or less the same. However certain areas are more privileged than others; these are usually the more strategic markets which are better served so the issue of imbalance in coverage remains. Indeed, efforts to translate the news for grassroots audiences do not resolve the challenge of imbalance in service. Broadcast news translation is complex, requiring an interpretation of a text that is written (albeit for oral delivery) to a language that is culturally distant from the source language. Translations are prepared within time constraints as the production schedules dictate, and the adequacy of the efforts depends on the competence of translators. These issues have been documented by Simpson (1985) with regard to the precursor for television news practices—radio news. Many of his observations remain valid. They are more acute in certain states that have a complex linguistic composition. In such states of the federation, like Plateau State, less familiar local languages are brought into limelight, yet instead of fostering balance in service, such attempts to reach out create new imbalances.

NTA Jos is the station that serves Plateau state. There are 63 languages spoken in this state. The station was involved in a pilot scheme to enhance the production of news in local languages. The effort was in response to a federal government directive passed through the directorate; that stations must communicate in more of the local languages. Since it still had to fit its local bulletin within the allotted 30-minute slot, NTA Jos had to select which of the many languages in its operational area to broadcast in, bearing in mind that it still had to produce the news in English and Hausa. The news bulletin is presented in three segments of 10 minutes each; one local language, then Hausa, then English. The decision was taken to choose five of the local languages, one for each weekday.

The local languages included in this pilot scheme are Taroh, Goemai, Birom, Angas Markafa. Each of these is featured on a particular day of the week for 10 minutes. This is hardly adequate to do justice to the issues of each of those communities as well as the issues of general interest. Meanwhile, other communities have been alienated because their languages are not featured at all. This exacerbates a different sort of imbalance.

In spite of the effort to reach them, feedback shows that sections of the audience are not satisfied with the policy. The serious ethnic rivalry in the state may be responsible for this. For instance, arriving at the choice of languages for the scheme was itself a struggle. Though the station sought guidance from the state House of Assembly, it got none. The representatives of the people could not agree on which five languages should be regarded as most prominent. It is not possible to feature all languages, though for these groups to be involved in the development process that is being orchestrated around them, they need to be involved.

Having recognised the limitations of vertical (top-down) models of communication, scholars like McQuail (2005, 1983) had favoured more participatory models, in which local communities would be involved in the identification of the needs and the solutions. Others, such as Moemeka, advocated that stations be closer to the people to foster their participation (1994: 61). This received wisdom is also held by practitioners in the field, including United Nations agencies such as UNICEF. Enemaku sums up the argument for participation in development-oriented communication succinctly when he argues:

> . . . it is urgent to shift from posting messages to [having] real dialogue . . . from problem-identification to solution-encouragement and appreciation. There are many good things that individuals and communities are doing, and development programming should both appreciate and promote such positive developments. There is also a need to shift from expert solutions to community solutions. (Enemaku, 2006: 56 – 7)

The News Directorate and departments are strategically positioned to facilitate these. However, for the participatory model to work through them, they have to overcome challenges of manpower and logistics. The task requires local stations to have news teams who are competent in the local languages and cultures. That is not yet the case, so in spite of the proliferation of stations, many are still far from their audiences. The distance could be both physical and cultural. As demonstrated by Esan (1993), audiences regard even locally produced programmes as foreign when the language of broadcast is incomprehensible (whether English or a local language). Such programmes are culturally irrelevant. Stations which operate in erstwhile regional centres such as Jos are thus distant from some of the final audiences when practitioners are not indigenes. Producing bulletins in local languages was a task for which those at the station were ill-prepared. It meant they needed to work through translators, since they did not speak the local languages.

The station relied on the communities to provide translators. These are given only a modest honorarium. Because of the embargo on recruitment, translators could not be employed This is hardly ideal for the requisite controls in the news production process, as a Manager of News lamented.

> I do not understand any of these languages, and what they say when the time for news comes, we do not know. They may decide to insult the Emir [paramount traditional ruler], and if they say it in their language, you know it is dangerous . . . I can't vet what they have written. If you say that government means to pay, and they go on air

to say government has said it won't pay . . . and there is a riot on the streets . . .

(Senior News Officer, 2008 [identity withheld],)

Such subversive potential in the system would be easier to control if the production team and supervising producer were competent in the language. To expect that professionals will have such varied lingual competencies is unrealistic. They could train local volunteers in the rudiments of news production, though the threats of misplaced loyalties remain. Herein lies the wisdom of community stations. Jos may no longer need to transmit in Taroh when the Langtang station, which is situated among the people, becomes fully operational. When the community stations are fully functional, they could close the gap between government and the people, and foster development of these areas, assuming that the good intentions are well executed. Clearly the News Directorate is saddled with constraints that reflect the upheavals in the network as an entity and the need to give adequate coverage to a nation that is now made up of 36 states.

The political rivalries, distrust and the mutual insecurity of the various leaders placed constant pressures on the operations of newsrooms across the nation, including the News Directorate. At the inception of NTA network operations, and right through the 1980s, network news had a 30-minute slot. This was meant to feature news from the 19 states that existed in the federation. The increase in time for the network newscast from 30 minutes to 45 minutes offered an increased opportunity to cover news from around the world and the states. However, rural news could hardly compete with news from industry and the commercial sector, sports and so on. Logistics for news coverage in the other sectors were considerably better, since these were higher on the scale of priorities. News, particularly as defined for television audiences, was urban-oriented. Shotumbi, once a Deputy Director of News, acknowledged the predominance of urban issues and events in news reports and discussions.

When rural areas are featured, it is most often in relation to some function performed by a governor, minister or local-government chairman. The rural people themselves live and die undisturbed by

television cameras unless they are involved in communal clashes or natural disasters.
(Shotumbi, 1996: 178)

This pattern of reporting is reminiscent of the skewed global information order but the organisation has since acquired the capacity to reach out more through grassroots-oriented programmes, especially in its collaboration with the Programme Directorate's initiative, AM Express. This is a real achievement, in that news is no longer being disseminated from one central urban location. It reflects one of the gains in engineering, since such coverage has been facilitated by the acquisition of the new DSNG vans. The following account illustrates some of these gains.

Q: Can you describe the previous pattern of news coverage?

A: In terms of news, all the NTA stations were part of Lagos [when it was the operational headquarters and network transmission was anchored there]. You have a report in Kaduna here; that report can be compiled here [but it has to be] sent down to Lagos. The tape is sent down to Lagos usually by road. You will assume that if it were sent by air, that the story may have a chance to make the bulletin on the same day, *for where!* [Colloquialism meaning 'not a chance.'] If you take it by road, it will take you one or two days to get there. News is already about two or three days late before it is broadcast. But now, we thank God, we can just fax our stories to headquarters. What is happening today, by 9 o'clock you can actually see it on the news. These times have more advantages than the past . . . NTA coverage in [the rural areas] is now far much better because . . . we are creating more stations, unlike before. In Kaduna state we had only one station [in Kaduna City itself] but now we have NTA Birnin Gwari, NTA Kafanchan, NTA Zaria, just to enable us boost information to the rural areas, so you can see we have more advantages. The coverage is very wide now.

Q: Besides technology what else is responsible for the improvement?

A: In terms of live coverage, during the Java celebrations [an annual event that happens around Easter, held in Southern Kaduna, about 178km from Kaduna metropolis] prior to now, we had to go there to package the story, the entire event is recorded and brought to the station. It may be used as an item in the news. Now, because of the extension of the network broadcasting, you can go there and transmit and the audience will get [live] whatever is happening in that place.

It is heartening to see that Nigerian television is offering what it had regarded as ideal news coverage across the nation. The technology at its disposal and the administrative structures prevented it from realising the ideal, but these are changing. With the creation of community stations, new administrative demarcations that make for more effective decentralisation of service have been outlined. In this we see that differences in language highlight ethnic and sectional interests, but other issues that were not so readily apparent determined the relevance of news. Available technology was an issue, as was the available time.

When the television day was shorter, news was a constant feature of the schedule. In 1986, *Network News at Nine* was the flagship of the News Directorate. This was a 30-minute bulletin aired between 9pm and 9.30pm. News in Brief was featured at 5pm daily for five minutes. Newscap featured at 11pm for 15 minutes. Back then, local stations also had the 7pm slot for their local news. In some states, such as NTA Kano, the entire hour between 7pm and 8pm was given over to news, going by schedules from 1979-1982; the first half hour in English and the other for *Labarai* news in Hausa

By 2007, 24-hour transmission had been introduced. The schedule from the Directorate of Programmes shows more opportunities for disseminating news. *Network News at Nine* is now one hour long. On Sundays there is an extended slot for *Newsline*. On Wednesday, there is the midweek *NTA News Extra*. There are slots for News & Current Affairs programmes at other times of the day. For example, *Panorama, One o'clock Live; One on One, Nationwide*. In addition to these, there are sports programmes, business news, political programmes (*Inside the Senate; You and Your Reps*) and other specialist beats. Some of these are mandatory network programmes. Others are optional feeds.

Fig 6.5: Example of Network Schedule Third Quarter 2008
Courtesy NTA Programmes Directorate

Fig 6.6: Example of budding network operations -
station schedule (NTA Kano) Third Quarter 1979

Without the extra time made available through the introduction of morning transmission so many opportunities for featuring more areas of the country on such a regular basis might not have arisen. *AM Express* offers a platform for newscast. It is scheduled in the same time slot daily through out the week. There is a rota to take contributions from all (zonal) network centres. Each network centre plans which of the stations within the zone will contribute to the day's transmission. This increases the chances of the representation of diverse areas on the network. This organised decentralisation of effort, along with the improvements in equipment, helps the directorate make an impact. The improved coverage of remote areas within Nigeria and its diverse peoples is a foretaste of what could be achieved on the African continent.

The News Directorate was once confronted with a shortage of reliable news-gathering and processing equipment. This led to an over-reliance on talking heads, TV news reports were stories that audience heard but did not see. This was worsened by the sight of newsreaders with heads buried in their scripts. This is no more the case. It is now usual for newscasters to make eye contact with the audience as they rely more on autocue cameras. This technology has played a vital part in enhancing the appeal of news, but there was much more to this improvement than technology. Success of news operations depends on the right logistics, for instance being at the events at the right time which calls for reliable vehicles, reasonable drivers; reports require adequate footage that is tightly edited, sometimes archival material are required. All these need coordination of personnel from different departments.

Fig 6.7: Autocue camera in use during newscast at Lagos Television

The Editorial Board

The following is an account of an observational study conducted in 1991. From all indications, the practice remains in principle, though it was not possible to establish a comparison by direct observation during the 2008 study. On a day-to-day basis, the Director of News chairs the Editorial Board. The Director of News is assisted by a Deputy Director, who in turn is assisted by Assistant Directors. Working with them are Managers who in turn were assisted by Controllers. There are Principal Officers and other Senior Reporters and Editors within the team. The Directors and Managers (occasionally Controllers) make up the Editorial Board. Each unit responsible for making programmes is represented on the Board. This includes the Programme Producers (Current Affairs), News Producers, the Reportorial Corps, Sports News Team, News Directors and those responsible for policy and supervision; that is, the Director News or a representative (either the Deputy Director or Assistant Director News).

Except at weekends, this Board meets daily to evaluate the previous day's output, and plan the new day's news. It sometimes discusses in detail the strategy to be employed in the treatment of certain news items, although this is the responsibility of individual producers and reporters. The depth of deliberation may be informed by how sensitive or controversial the story is adjudged to be. The more senior members of the board give a steer in such instances. When necessary, they get legal guidance or seek further clarification. Since the Director General is the Editor in Chief, the buck stops at his desk. A similar hierarchy occurs at stations at the state level, with the General Manager as chief executive and the Manager News as deputy, with Controller and Principal Officers from the relevant units represented on the board. Though it is unusual that they are involved in great detail on a day-to-day basis, there is good reason for chief executives to be at least informed of what tends to be the most sensitive output of their stations. Their personal intervention makes a difference at critical times when there is trouble with the authorities regarding the news. CEOs have to protect their staff and defend the integrity and independence of their station, sometimes by negotiating the reprisals that accompany offending news stories. Therefore it makes perfect sense that they keep close tabs on the News Directorate (or Department).

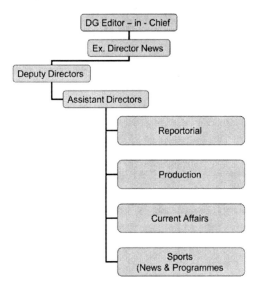

Fig 6:8: Organisation of the Editorial Board

The Newsroom

This is the hub of the activities of the News Directorate, which originally was the only Directorate designed to produce programmes. A corps of reporters is deployed to gather news. They were organised around specific desks to encourage specialisation that will foster better rapport with audiences and contacts on the beats. There are correspondents for State House, the House of Assembly, Education, Health, Science and Technology and so on. The organisation of the desks has been subject to change as determined by priorities. For instance, once the NTA headquarters had moved to Abuja, there was no longer a need for an Abuja correspondent. Directors will also have their own interpretation of the job, experimenting with different arrangements with their staff. The merits of specialisation have been appreciated consistently over the years. Desks or assignments are still regarded by the specialty, theme or sector to be covered.

Perhaps because of its size or the sensitive nature of inter-cultural relations, the NTA still attempts to ensure order and fairness in the devolution of duties. Different tactics had been explored over the years. In news coverage, attempts have always been made to identify areas of jurisdiction in the NTA structure, between reporters at the network headquarters and those at the local stations. The plan was to maintain a balance between better equipped, more prominent and more experienced reporters and the fledgling teams at the local stations. Local teams are expected to be experts on subjects in their backyards; they are also expected to develop professionally. Opportunities for showcasing their work to network audiences may just be their professional break. This had been one way that new talent was discovered. So, in pursuing stories, reporters have to keep to their territories. Abuja- or Lagos-based reporters should not initiate stories outside their territory. Instead, they should commission another reporter resident in the area of interest to cover the story on their behalf, since these are supposed to have greater local knowledge. This may be frustrating for reporters on specialist beats who may not have local reporters who are sufficiently aware of the issues at stake. Therefore, laudable as the motives may be, these policies need to be accompanied by reasonable mentoring schemes that will enable the more experienced reporters to enhance the effort of local or grassroots reporters through such collaborative productions. If, because of poorer skills and facilities in the local stations, the story is substandard, there will be frustrations all round – for the reporters, the audience and the organisation. This is the sort of dilemma that global news organisations have to contend with in their attempt to reverse the skewed information order.

Challenges of News Production

Of the challenges encountered during news production, some can be traced to the context of production, while others are inherent in the professional orientation and interpretation of the duties involved in news service.

There is no rigid approach to news-gathering, but the underlying interests that informed the setting up of the organisation are not overlooked. The NTA Code spells out the responsibilities which guide the operation of the directorate. As part of its social objectives, the NTA is required to

> bring news and information to as many Nigerians as it can physically reach. Such information must be factual, accurate, impartial and relevant, thus serving as a means of arousing in the people, common emotions and directing their attention to specific national objectives. (NTA Code 1981: 31)

From the political standpoint, the imperative is to maintain national cohesion, promote political awareness and strengthen the unity in a free and democratic society. There were technological and cultural objectives that also had a bearing on the definition of news. To this end, the directorate seeks to tackle items which audiences find interesting. It is geared towards enlarging viewers' understanding of themselves and the community in which they live. It also features the audiences' environment, their problems and aspirations.

The News Directorate claims its news programmes are meant for everybody—the entire Nigerian populace—people in business; in rural areas; women and special-interest groups. All stations are mandated to hook up to the network news at 9pm. The directorate recognises that its market is wide and varied and it believes that everyone must be served.

The directorate aims to inform, to create awareness, and stem the tide of undesirable social traits. It is supposed to offer service that has a wide appeal. In day-to-day practice, news items are juggled and "balanced" against operational and other interests. It has regular slots for business news on the Friday newscasts geared towards the business community, the specialised *Tonight at Nine* for the intelligentsia, and *Newsline* with its human-interest approach. Its Sunday-night slot targets a general audience. Other news programmes that have made their mark over the years include *News Extra, News Panorama* and a range of sports programmes.

The NTA has a sense of duty to preserve and protect the sovereignty, cohesion and unity of the nation. The paternalistic orientation of the NTA comes through in operational conventions especially in the coverage of politics and the assessment of aesthetic quality. Its coverage of an air crash, as discussed earlier (Chapter 5) is an example of its sense of social responsibility. The incident demonstrates the variation in the interpretation of the concept of responsibility and explains the variety of perspectives to which Nigerian television audiences are exposed in a competitive era. (Details of this have been highlighted in a separate chapter).

This following excerpt from a very senior news producer gives an insight into the interpretation of the policy and the professional orientation that exists in the directorate.

Q: When you say you try to include something for everybody, does that mean you try to include a news item for the different segments in the audience, or do you, in your treatment of each item. take into consideration the different segments?

A: No, you can't do that for each item because it might not relate to everybody. We try to include something for everybody and from everybody. You know, proper communication is two-way, should be a two-way thing, so whilst we are carrying information for people in the rural area, we also bring things from the rural areas; to remind them or let them know what is going on; whether it be government level or on the very personal level of some festival in some village. That the man in Maiduguri [North-East] may now know about what is taking place in Owerri [East].

Q: But in terms of priority, if you had a choice . . . ?

A: It will be that which is more national, something that affects most people. We take it [the news] from the national down like that [to the states, local governments, wards and specific communities].

The admission suggests that news, far from being meant for everybody, could be divisive in subtle ways. For example, in its processes of selection and treatment of stories and issues, it outlines the boundaries of the audiences to be served. Another observation from the editorial meeting was the heavy reliance on specialists to discuss, explain or predict patterns. The need for this practice was explained as follows.

> Because broadcasters are not supposed to hold opinions, they need to find resource persons who can articulate their thoughts. They [broadcasters] are acknowledged as jack-of-all trades [but] by inviting 'resource persons' they accept that these ones are more knowledgeable and so better qualified to explain or predict events.
> (Interview conducted in 1991)

This demonstrates an acknowledgement of the limits to the knowledge of journalist, though it does not address the limitations imposed by the need for speed in routines of producing broadcast news. "Resource persons" tend to be people of stature in their field of endeavour; people who can be easily recognised by the public. Their appearance on TV is therefore designed to lend credibility to the point being made. It may be, however, that their clout overshadows the issue that they are invited to address. Another quality that is considered in selecting a resource person is the ability to express himself or herself on television. In all of this we see the dynamic tension of balancing credibility and clarity. However, it also shows how news could become stilted in its discourse.

Relying on the restricted pool of resource persons makes the news predictable. It also gives it an air of exclusivity, decried in certain quarters as evidence of sloppy practice. The pool of resource persons is often limited to particular academics, especially those who can be easily accessed from the television station. This is understandable in the event that experts have to be brought into the studio for recordings.

However, given the timeframe in which stories have to be turned around, and the other pressures that television journalists have to contend with, one begins to understand why it happens. With technology that makes recordings and live transmission more flexible and moveable, that reasoning will no longer be tenable.

Besides the impact of routines that mould newsroom practices, the directorate deliberately eliminates the consideration of certain audience categories from aspects of its service. This is based on the assumption that certain programmes are pitched above the level of particular audience groups; they simply cannot understand, nor can they contribute meaningfully to those debates. This unapologetic stance reserves discussions about serious politics (governance) and economics at global, national and local levels for the elite. The directorate is able to do this if it expects the local stations to provide such opportunities for the masses.

According to the Deputy Director News in 1991,

> The audience of news in the generic sense is anyone, from policy-maker to peasant farmer to local artisan; yet there are a number of sub-audiences within the large audience. The largest and most significant is, of course, the elite—the middle- and high-income group . . . So, while news is designed for the entire audience, the elite and non-elite, Current Affairs programmes [are] for the more politically inclined section of the population. Those who are [most] interested in knowing about national issues tend to be [those] concerned about [the impact of] policies on the group to which they belong, as individuals, on their businesses and [as] members of the labour unions or bureaucrats.
>
> "So current affairs programmes are designed for that category of audience, and they tend to be professionals, businesspersons, members of the labour unions, bureaucrats, students and so on. "They are the kind of people we design Current Affairs programmes for. News [, however,] is for the generality of the audience who have access to television.
>
> (Interview conducted in 1991)

Such positions are often justified by research findings but sometimes the authority relies on anecdotal evidence, feedback gleaned through the interaction between staff, in their daily lives and the course of their duties, and members of the public. Such evidence is used to shed more light on limited scientific research. When planning is premised on this evidence, there is a risk of being misled and of poor representation persisting. When there is no-one to speak for them, audiences on the fringes will remain excluded.

The directorate also promotes government policy by attempting to foster appreciation for government plans while strenuously denying that it is no more than the public relations arm of the government.

According to the Deputy Director of News cited above,

> News is not uncritically reported . . . government policy is aired at times. However, there are occasions . . . when there is a clash, conflicting demands of various powerful interest groups, and this must be and is considered.

(Interview conducted in 1991)

This reality is an acknowledged source of disillusionment for young reporters who find it hard to compromise between what is professional and the concessions perceived to be required by proprietors. Similarly, the poor reputation of NTA News being a government mouthpiece, an organisation which fails to maintain professional standards, is a concern for those who were professionally minded in the organisation.

From inception, NTA mirrored the national priorities. Its mission is "to provide excellent television services worldwide, and project the true African perspective." This is a restatement of an objective that is as old as the organisation itself. The NTA had always desired to be the mouthpiece for Africa, in line with Nigeria's strategic position on the African continent and in the ECOWAS sub-region.

Some speakers at the symposium organised by the Obafemi Awolowo Foundation, marking the 40th anniversary of television in Nigeria, argued that television journalism should be relevant to the social needs of the people. Having painted a dismal picture of the Nigerian state, including issues such as shortage of housing, high population density, poverty, hunger, low agricultural yield, erratic power supply, bad roads and other factors that undermine the prosperity of the people, Shotumbi (1999) concluded that television producers and journalists should help to create an awareness of the magnitude of the challenges at all levels of government. In his paper, Akinfeleye (1999) advocated that the medium's potential for entertainment and information be harnessed to mobilise citizens for social development. He also advocated the establishment of community television stations.

The communiqué adopted at the end of the symposium observed that television was a stimulus for development and must be used to play its role in promoting the development and prosperity of the people. Television had been used to promote the best in Nigeria and to show the world the strength of the continent; that Africa has more than the familiar beggarly, donor-dependent image. These views have since been codified in the organisation's mission statement and they are well known among the staff.

According to one of the reporters interviewed in 2008,

> We [the NTA] have the strength [capacity] to be the African CNN, if only we had the independence that they enjoy, their exposure and the capacity to be on site [respond to incidents] as they occur. They [CNN] have good archival material [and their] capacity for research is wonderful, so you can work on your stories quickly and have a comprehensive report, but here you struggle with one story, then you have to be on a queue to edit . . .

These frustrations take the edge off the pleasure that the job brings for this Nigerian reporter. In spite of the limitations of her organisation, she is aware that her output is judged by the standards of trans-national media organisations. It is customary for managers in TV stations to monitor the output of their station and that of competitors. In several instances the monitors were tuned to these trans-national media corporations—CNN, the BBC, Al Jazeera— occasionally they were also seen monitoring the local competition. This is evidence of an external pressure imposed on the News Directorate. Yet staff, like the reporter quoted above, appreciate the crucial role they play, regarding themselves as crusaders and agents of change. For them, every chance to make an impact on social life is rewarding, making the effort worthwhile, as articulated in the following excerpts:

> We are involved in civil advocacy; we are involved with the core issues that affect our lives, whether [they be] HIV/AIDS, voter-education (not just campaigns and elections) spiritual upliftment or tradition.

We are a cultural medium — we use the power of radio and television to project cultures and traditions to people across boundaries, who then appreciate us for what we are. So anyone who finds himself doing this job should see himself as privileged beyond the professional call."

(Ladan Salihu, former NTA News staff, Zonal Director, Federal Radio Corporation Nigeria - FRCN 2008)

What keeps me going is the beat I am on, that is to promote preventable diseases. With hindsight, 60-70 per cent of the diseases ravaging us are preventable. I am involved in human development and see myself as a positive change-agent.

(Moji Makanjuola, Senior Reporter NTA News 2008)

For someone like myself who has multiple influences in his background (one who comes from a royal family, has an Islamic religious background, and is a religious scholar) relying especially on the point of view of a scholar, my orientation to work is [to offer] total submission to the will of the authority that employs you. [One should give] total commitment to the will and dictates of your employer. Then you do those things that further the cause of the nation. But the system has changed completely. Some do not want to learn. They do not want to know how the job is [to be] done.

(Adamu Abdulahi, Senior Reporter NTA News 2008)

Constraints identified in the performance of the directorate are imposed by the poor state of social infrastructure in the areas where the organisation operates. A lack of telephone services, efficient transport system, reliable electricity supply and adequate library services are all realities that television journalists have to live with. These are made worse by funding constraints.

Those obstacles have provided ready excuses for those who want to be slack. Given recent investment in camera chains and editing facilities, there is less basis for excusing inadequate footage in news reports. In any case, more industrious reporters would make no excuses for poorly researched reports, inadequate use of archival footage, poor visual consideration and so on. They find ways round these hurdles.

As discussed earlier, NTA had to resort to the commercialisation of news and other "aggressive" revenue-generating drives first seen on state government stations, despite its enviable position as the sole network in the nation. Commercial news refers to items which are identified as having commercial value to the subject or sponsor of the report. If certain items are not carried except the subject pays the assigned rates, some genuine news items could be at risk. Others more deserving of attention may be sidelines because they do not benefit the station. Unlike newspaper pages, broadcast time is inelastic so the choice of one story puts another at risk. Consequently news may become the voice of those who can afford to pay—government agencies, big businesses and rich individuals.

The competence of NTA journalists is sometimes judged by the output from such a challenging operational environment. Yet their record when they find themselves in other local working environments, such as those in private television stations, or globally, like Yusuf Jibo on CNN, confirms their professional competencies. Nigerian professionals can compete with their colleagues in other parts of the world. News operations betray the foibles of the individuals involved in the news process, but the directorate of news also reveals the enormity and complexity of the task of keeping a nation as large and diverse as Nigeria informed. It is a Herculean challenge. The organisation has the best intentions. However, because there is often so much at stake, there are no guarantees that all the players in the process have the best intentions of the nation or others at heart.

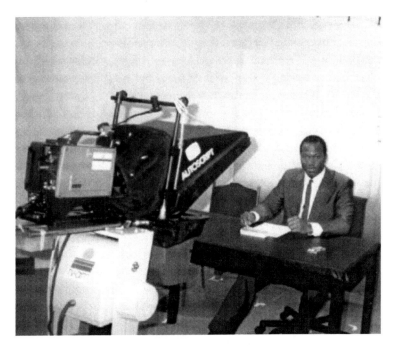

*Fig 6.9: Foreign Exposure for a Nigerian newscaster
(Courtesy John Momoh in TV-AM Studio, Wales)*

The Programmes Directorate

Besides the news, the output of the Programmes Directorate is perhaps the most visible in the organisation. This directorate is responsible for all the non-news programming activities of the authority. An examination of its orientation, organisation and activities is essential for an appreciation of the output of NTA stations. This is the focus in this section. The Directorate of Programmes has had to respond to emerging priorities within the organisation, as well as changes imposed by the wider society. Nonetheless, the structure in the directorate has been informed by the organisation's objectives. There are social, cultural, economic, political and technological objectives set for television broadcasting. These have been summarised as follows

I. To encourage the preservation and development of humane values and respect for the dignity of man.

II. To promote social justice based on responsibilities and rights of the individual in society.

III. To promote the development of a high level of intellectual and artistic creativity.

Because this directorate is responsible for the bulk of the NTA's programming, it may well be regarded as the design and quality-control centre. The activities of the Programmes Directorate are therefore indispensable. It is responsible for planning the schedule for network programmes, and determining how much space is available for local programmes. Network programmes have accounted for the lion's share of programming on each NTA station from its inception. In 1991 there were 12.5 hours of network programming weekly. About 41.6 per cent of these programmes were featured during prime time, between 8pm and 9pm. The programme schedule for the last quarter of 2007 shows that the Programmes Directorate catered for a 24-hour broadcast day. On most days, only two and a half hours were clearly marked out for local programmes; a half hour each at 6am, 2pm, 5pm, 7,30pm and 10pm. Just one of these slots (7.30pm) can be regarded as a reasonable time of day considering the lifestyles of the audiences, which include children.

Even when programmes were not featured on these plum slots, they were strategically scheduled to reach their intended audience at the most opportune times. In 1991 the 6.30-7pm slot on weekdays was reserved for children and youth audiences. There were five 30-minute programmes in a week for this age group. Formal teaching sessions through *Instructional Television* was on the NTA schedule at 4.30-5pm. and stripped across the five weekdays. Although *Instructional Television* was no longer on the schedule in 2007, there were still educational programmes such as *Work it Out, Take a Step, Reader's Club* and *Fun Time* on the schedule. These are products of the NTA Educational Television service.

Children's programmes have since been restricted to the 5-5.30

pm slot, except where the local station decides to provide different slots. In the third quarter of 2008 there were only two programmes, *Talking with Youths* and *Tales by Moonlight*, in the children's belt. Young people had to wait till Saturday or Sunday morning if NTA stations were all they relied on. Even then, they would not exactly be inundated with programmes to choose from. It seems the NTA has conceded that this group is better served elsewhere in the competitive era.

Programmes that have taken over the children's belt (between 6pm and 7pm) on weekdays are broadly enlightenment-oriented dramas and other shows. Though not particularly designed for children or young people, the themes on those programmes such as *The Journey so Far, Consumer Watch, Corruption Must Go, Human Rights Platform and You and The Economy* commend them to younger viewers in the hope that they will be more politically aware and better tuned in to their rights and civic duties. Yet if the programmes are not sufficiently sensitive to these critical youth audience segments, there is a danger that this use of time may alienate them. A number of these programmes are efforts of independent producers, though the directorate oversees their work.

The Programmes Directorate has the widest-reaching tentacles because it is the one that truly engages with all categories of the audience; reaching young and old and cutting across social demarcations. Its structure has also been subject to changes over the years. Some of the upheaval was due to external interference as well as internal politics. One would have thought that positions in the Programmes Directorate unlike News Directorate were not as sensitive or subject to wrangling, but it has had its share of such drama.

It was initially conceived as a policy warehouse for NTA programming and headed by a woman; she was the only female Director in the authority at the time. Dr Victoria Ezeokoli was the Director of Programmes and she was appointed to bring intellectual direction to the operations of the directorate. According to the Director General when she was appointed, (Vincent Maduka) the directorate was not intended to be involved in the production of programmes. The post of Director of Programmes was meant for a person who was more of a critic; one who could look dispassionately at the output of the organisation and make recommendations. It was a job for an ideas person, one who would be able to inspire, motivate and curtail the excesses of the glamorous personalities within Programmes Departments.

If television was to have an impact on society, if the medium was to be a teacher, as was originally conceived, it required some intellectual stimulation. Among those considered for the post were Professor (then Dr) Bolanle Awe, Dr Akin Euba, Dr Laz Ekwueme and Frank Aig Imhokuede. None of these had worked in the organisation, though they had varying degrees of involvement with the arts, including television. It was therefore assumed that they would come with fresh perspectives. For various reasons they all turned down the offer. A career in television management may not have been prestigious enough compared with the paths that they were charting within academia. Perhaps this showed the low social esteem that working in television had in the early days.

Still, the Director General was convinced that the role was cut out for an academic; it did not require hands-on expertise in television production. Dr Ezeokoli was contemplating advancing her career in academia when she was considered for the post. She was interested in the job, she fitted the bill and she was hired. Not everyone in the organisation agreed with that decision. There were those who believed that it would impede their prospects for progression. They desired a Directorate of Programmes that had room at the helm for people with their expertise as programme-makers. There was internal politicking, and, as is typical in establishments, some whispered their concerns to powerful friends outside the organisation. Soon external parties began to mount pressure. Even though the Programmes Directorate was ostensibly less political than News, recruiting for the post of Director of Programmes created quite a stir. Politicising this issue led to changes in the structure of the directorate.

The introduction of a National Television Production Centre (NTPC) marked the beginning of programme production at headquarters. According to Maduka, the NTPC was the brainchild of Olu Adebanjo, the Special Adviser to the President on Information who was in effect the Sole Administrator of the NTA (because there was no Board of Governors constituted at the time). The NTPC was headed by the late Adamu Augie, who till then had been a Zonal Managing Director. He had been invited to Lagos in the first instance for a special assignment—the production of Verdict '79, a programme which offered analysis of political events in the lead-up to the elections for the Second Republic. He was offered other duties at the expiration of that task till he assumed office as head of NTPC.

With this precedent, staff in the NTA's Programmes Departments had a place in headquarters to aspire to. The initial idea of having a Programmes Directorate that did not rival its state stations in programme production was temporarily scuttled.

The Director of Programmes is assisted by a team of Deputy Directors and Assistant Directors. The hierarchy is typical of News and other Directorates. The Director of Programmes' assistants generally have specific responsibilities. In 1991 there were five lines of duty characterising the basic duties of a Programmes Department (in this case a directorate). The ranks have been fairly consistent, even if reporting lines now vary. The directorate consisted of four substantive departments, along with one for Special Projects.

Planning and Development

Planning and Development was the largest of these departments. It was responsible for scheduling network programmes, including those from the News Directorate. It had responsibility for regulating and monitoring policy. This it did through its divisions – Research, Documentation, Regulation and Planning. The **Research** division was expected to monitor the performance of the stations under the authority's umbrella. This was at a time when there was no external monitoring body such as the National Broadcasting Commission (NBC). This was a critical task but it was soon moribund.

> We used to have a proper monitoring outfit at the instance of the DP herself but its efforts were frustrated . . . We used to have Monitoring Officers from headquarters go into different towns unannounced, check into small hotels and monitor reception as viewers [watching what these see, not signals within the transmitting stations].
>
> After a period of about three days the officer would report to the station he had been monitoring, first to confirm that he was there, and then [present] his report. We used to care about even the

language and the technical quality of content; but we came up against so many obstacles.

We used to have the stations monitor themselves and send in monthly reports. We had standard monitoring paper [format] which I could have given you a sample of, but I doubt if it can be found anywhere now ... They were really broadsheets properly printed and could not have been missed. They were stacked in half a Portakabin and they disappeared. The people who moved into the office claimed they saw it, but they assumed somebody moved it into some store. We searched all the stores there was no such thing there.

Someone must have sold these to some groundnut seller because I can't think of what else was done with them ..."
(NTA Monitoring Officer interviewed in 1991)

The detailed account above is presented to illustrate the way that poor attitude to such a crucial job could undermine its success. This was the unit that was meant to ensure that the NTA adhered to the codes for programming.

We monitor[ed] before and after transmission, and the general purpose was to ensure that the programmes keep [to] the NTA objectives. They look for anything vulgar, indecent exposure, errors in transmission. [Erring] Stations were usually queried but nothing really worked. There were a million and one excuses. It was like a waste of time.
(NTA Monitoring Officer interviewed in 1991)

It became worse than a waste of time when Monitoring Officers realised that their effort made little difference to the job. They even began to experience antagonism at crucial periods in their careers. The research had been mistaken for a witch-hunt. In the light of the many frustrations experienced all round, the extra pressure was not welcome. Little wonder the system did not endure.

My involvement in the monitoring exercises affected me at promotion time. GMs had taken it personally that I had queried their station, so they ... had a chance to pay me back during the promotion interview – as soon as I had been asked what my previous duties were and I

included monitoring . . . [conversation trailed off]
(NTA Monitoring Officer interviewed in 1991)

This was how unpopular the operation was, irrespective of how effective it could have been. The NTA is not isolated in this ailment. There is still a great deal of insecurity and distrust, even in the wider society, where people take a dim view of critical comments. Research findings that were critical were not always regarded as constructive, regardless of the reasons for the research effort. Research then was meant to inform policy and was crucial at the time when there was no external regulatory agency.

Today the NBC monitors the NTA and other broadcasting organisations. The NTA's in-house monitoring effort has also been revived since 2007. Driven by the desire to attain truly international standards of operation, and to convince advertisers of the value for their ad-spend, the NTA's management approved the inauguration of a Monitoring Scheme. The team, headed by the Director of Training and Capacity Building, operates nationwide. From all accounts, the system is similar to that described above but there is an attempt to protect the identity of the monitoring officers by rotating them around the zones. With the tough talk about sanctions for erring stations, such protection is necessary if the experiences of the past are anything to go by. The situation may be even worse now, since General Managers are higher in the hierarchy than they used to be and wield more influence.

Programme Acquisition

The Department of Acquisition has also attracted criticism. It was responsible for acquiring programmes, both local and foreign, for the Authority. All programme procurement by members of the authority passed through this department.

In the 1980s the directorate invested much more in foreign programmes. Each station decided on its own selection and scheduled these as appropriate for its audiences within the approved limits for imported programmes. So NTA 2 Channel 5 had *Palmerstown U.S.A.*, *The Sweeney; Return of the Saint; Some Mothers Do 'Ave 'Em, Mixed Blessings* and many more in the first quarter of 1986. NTA Ibadan, as with other stations in the South-West, restricted its consumption of overseas programmes to children's programmes and the movies slots. The stations have come a long way since the 1960s, when foreign programmes were littered across the schedule to make up for the limited number of local productions.

These days there is a lot more syndication of locally produced programmes because more independent producers are working on a range of TV forms and a variety of themes. There are drama productions, talk shows, magazine productions that address a range of subjects and audiences. These include *Super Story, Who Wants To Be a Millionaire, Harmony Estate, My Three Sisters and Apprentice Africa*. As mentioned earlier, NTA Enterprise (another arm of the organisation) is now responsible for negotiating and acquiring rights for international sports programmes. In this, the directorate has returned to its initial remit.

At inception it was meant to facilitate the development of new programmes. Sometimes ideas are generated in-house, as staff may have ideas that they want to develop. In some cases the ideas come from external sources. These include people who have particular talents as well as some who are familiar with the workings of the industry, such as television or radio artistes. As educational provision in mass communication, theatre and the creative arts increased, people involved in those institutions also became involved in generating ideas. Some ideas even came from people who may be regarded as novices in the industry, but who had knowledge of the culture and audience and could spin a good story. Such programme ideas were presented to the authority sometimes formally and other times through informal discussions. It was only with time and greater enlightenment about intellectual property rights that the issue of remuneration became one of grave concern. Programme ideas were meant for either national or local audiences, although those for local audiences were usually sent to the states for further development. Many notable network programmes from the 1980s were developed in this manner. It is reported that *Behind the Clouds* began as an idea presented to the authority by a teenager, fresh from high school, awaiting admission for further education. If the programme commissioning process is formalised and made more transparent at all levels, it may promote greater involvement of people with raw talent, as well as those who have relevant academic experience but who lack the resources to launch out as independent producers.

Programme ideas also came from independent producers,. Some of them were former NTA staff who had resigned or been made redundant in the rationalisation exercise of the 1980s. Amaka Igwe, Zeb Ejiro, Fred Amata, Lola Fani-Kayode, Tade Ogidan and Tunji Bamishigbin continue to contribute to the industry after their stint with the NTA. The list would be much longer, especially if one were to reckon staff from the first generation of television stations and those like Tunde Kelani and Jimi Odumosu who moved to state-government or private-stations. The Independent Television Producers Association of Nigeria, which was registered in 1992 as a non-governmental organisation, is an umbrella body for such people. It offers a forum for reflection on standards and training, and a place for different generations of television producers to rub minds. Through this forum, they continue to generate programming ideas to develop television and the Nollywood industry. The quarterly meeting of all NTA Manager Programmes to assess performance is another forum where ideas are generated, before they are developed into programmes. This meeting is also used to reinforce guidelines on programmes or station policies.

Library & Documentation

Library & Documentation was responsible for storing programmes and for copying and distributing centrally acquired programmes. This was no mean feat, considering the number of stations within the authority and the number of programmes and episodes which had to be dubbed and distributed. In some cases, because of the ineffective network transmission, programmes—which ought to be received via satellite—had to be dubbed from headquarters and sent physically to the stations. This meant that certain stations transmitted episodes of some scheduled shows a week later than stations that were closer to Lagos, which was then the centre of operations. Those stations that were nearer the centre had better reception of network transmission. Being behind other states gave a literal meaning to being described as a far-flung state.

The job of the library was not made easier by the state of its equipment. The VCRs and editing suites in the library were inadequate for the demands placed on them. The library was housed in Portakabins because of the space constraints experienced in Lagos before the move of NTA Directorates to Abuja. It is little wonder that the library could not offer better services or be more reliable. With the investment in more efficient transmitting facilities and the upgrading of equipment in the stations, some of these problems have been addressed but there are still challenges with the library in its more conventional uses.

Television stations, whether at headquarters or in the states, lacked a good book or documentation library in 1991. The situation is no different even in 2008. This is discouraging, given the importance of research in routine productions. A division that is responsible for supporting the research and scripting of programmes and concerned about accuracy and standards should be equipped for this task.

Similarly, the NTA has not done well in archiving its old programmes. A lot of productions in the early days were live shows. Even when recording began on videotapes, the mode of storage was cumbersome, taking up valuable space that the authority at headquarters and in many stations did not have. In any case, the humid climate in Lagos, location of the headquarters, was not conducive to storing tapes. Tapes need to be in cool, dust-free rooms, which meant tape libraries had to be air-conditioned. This was an additional logistical challenge. The conditions were not always met, as the value of archival materials was not always fully appreciated in the face of other pressing needs. The fact that tapes were expensive meant that a lot of programmes, especially in the states, were wiped, so that tapes could be reused. Producers had to guard their tapes jealously lest some unscrupulous colleague wiped programmes that had not been aired! Stories are told of archival tapes simply vanishing. Such disappearance cannot be regarded as innocent errors, when copies of programmes turned up as home videos in markets outside the country, notably along the West African subcontinent. In this way the directorate— indeed the industry—has been deprived of valuable assets; the legacy of archival material which could have formed the core of programming for new channels spawned with the deregulation of broadcasting.

Production Services is responsible for the design and construction of sets, props, graphics, editing and other equipment for post-production. Though equipment maintenance is the responsibility of engineering, user departments are responsible for using them. Production Services takes charge in this way, and its job is critical to the visual considerations of the productions. To facilitate the construction of large sets, the department relies on carpentry, painting and other such artisan-based trades, yet it was graphics artists who made the biggest difference in programming. Caption cards, backdrops, news sets were prepared by them. With improved technology, however, there is a drift towards digital composition of images.

At least Production Services is resident within the Programmes Department. This is great for planning and control, unlike the situation with other units that the directorate has to work with. Like the News Directorate, the Programmes Directorate is serviced by staff from other departments; Engineering and Technical, Marketing Services (Commercials), Administration and Finance (Accounting), for example. There are technical officers responsible for audio, lighting and videotape recorders (VTRs) and engineers responsible for the equipment and transmission, the administrative officers responsible for the staff team, including the drivers whose services are vital for location recordings, and accountants who must keep track of the budgets for production and ensure that the cash flow sustains the projects on hand. That these are all resident in their respective departments makes for effective control, though at times it constitutes a logistics nightmare. Situating Production Services within the Programmes Directorate suggests that, in principle, it is regarded as an essential element in the creative process. In practice with planning and budgeting, it is not always the case.

Fig 6.10: NTA Enugu graphics room display
Courtesy Peter Achebe NTA Enugu

Projects (& Social Mobilisation)

In 1991 this department was responsible for organising workshops, seminars and the National Television Festivals which also served as fora for identifying ideas and talent. Because these events acknowledged talent, they were also motivational. They were sources of programme ideas, and for the duration of the events, the productions were included in the programming. Many regular programmes made their debut on screen this way, as specials from this type of forum. For the period of the festivals, the normal programme schedule is suspended to give way to these specials.

The NTA held festivals that were exclusive to its own stations, and as the leader in the industry, it also organised activities that were open to other stations outside its authority. The NTA has thus been supportive of independent producers which it helped to nurture. These activities are seen as encouraging creativity and promoting excellence. It is certainly an effort at developing the television industry as a whole. As the media industry expands, many more independent parties, including those from the print media, now organise award ceremonies.

The roles defined for the Projects Unit made it subject to change. The Assistant Director (Projects) seemed to be more of a liaison and contractual-relations link between the government and international contacts. When programmes had to be entered for international competitions, these duties were essential. However, they overlapped with others in the authority. As the organisation evolved, changes in procedures meant that such duties were realigned with positions in other directorates like Administration. Likewise, attention to Social Mobilisation reflected the nation's emphasis at the given time. The Public Enlightenment Division in the federal Ministry of Information should be the body to promote social mobilisation assisted by the National Orientation Agency but successive governments still felt the need to establish specific social mobilisation campaigns. Notable mobilisation efforts in the past include campaigns for Expanded Programme on Immunisation; War Against Indiscipline; War Against Drug Abuse and so on. Recent initiatives include the 7 – Point agenda of the Yar'Adua Administration; the Nigeria Project; and the Rebranding of Nigeria Exercise. Though it may have been necessary to dedicate attention to these as projects, television is strategically positioned and tasked to address these issues on a systematic basis through its enlightenment programmes, nullifying further need for a social mobilisation unit. Engaging audiences in dialogues on a routine basis should promote long lasting change in attitude rather than short lived changes that a blitz of messages may generate.

Programming Units

Public Enlightenment and Entertainment are standard features in the Programmes Department of any Nigerian television organisation. This may be a legacy of the interpretation of the public service ethos —to inform, educate and entertain. Through the operations of these two units, a station's attempt to balance the duties of television as a teacher and as a medium of entertainment is evident. In the expression of its duties, the News and Current Affairs Department tends to be more concerned with promoting enlightenment, leaving the need for a distinct Enlightenment Department.

Public Enlightenment covers the routine production of documentaries, panel discussions and special coverage, including the broadcast of Grade A programmes. These are Presidential Addresses to the Nation, Budget Speeches, National Day Parades—all the programmes that must be aired. The department's schedule of duties included the planning and production of instructional television programmes; (less formally structured) educational fillers; jingles; and public service messages. There are striking similarities between its functions and those of the Public Enlightenment Unit of the Ministry of Information, or even the federal Department of Mass Mobilisation for Economic Recovery and Social Justice or the National Orientation Agency. Enlightenment programmes require liaison with other government agencies and should be proactive in identifying the aspects of living, worthy of attention. For example, sexual behaviour and reproductive rights issues have remained relevant because of their implications for population, development and poverty alleviation. In recent times, a renewed urgency to address these issues was brought about by the need to control the spread of HIV/ AIDS. In the task of responding to such imminent dangers in society, television should lead rather than follow a government agenda. In this regard, the

Programmes department (like News and Current Affairs) serve as sentinels in society.

The Entertainment Department is organised along the areas of responsibilities within it. Typically these are drama productions and features; music and light entertainment (quizzes, sketches); children's, youth and women's programmes. The department operates on the principle that there must be a proper mix of programmes and they must have a national appeal. Even the musicals played through the directorates' programmes must have a national appeal and bind the nation together.

Considering the sheer amount of time devoted to Network Programmes, the need for a broad range of programmes becomes apparent. This is all part of the strategy employed to ensure that NTA objectives can be met and that the organisation remains commercially viable by attracting audiences. Network programmes are drawn from a pool of programmes produced in local NTA stations around the nation. For instance, *Tales By Moonlight*, one of the most enduring network programmes, is a contribution from NTA Enugu and others. As a national programme, it has to maintain the national focus. Officers in the directorate ensure that programmes keep to their brief.

In presenting the Federal Character in its programming, the NTA reasons that it fosters the feeling of acceptance, a sense of belonging to every Nigerian, irrespective of where he or she is situated. By so doing, it assumes that unity is fostered. This reasoning takes into consideration sectional affinities. So different groups proudly identify with the programmes from their area, but others may also appreciate these shows. In this way the NTA raises the profile of groups which might otherwise remain "insignificant". In time, such exposure should broaden interests and foster appreciation and tolerance of other cultures.

Laudable as this is, the issue of how to ensure that the programmes appeal to the viewers remains. There are several reasons why a programme may fail to appeal to the audience even on an aesthetic level. The format and treatment of the subject matter must be carefully considered, whether programmes are factual or fictional. They must be entertaining and accurate. They must command the respect of the audience, otherwise groups may not be proud to identify with programmes from their area. Indeed, it may be that stereotypes about groups are reinforced because of lapses within programmes that are associated with them. The task of the Programmes Department is thus delicate yet essential for nation-building. How satisfied are the various audience groups? Or is the attempt to satisfy so many people and groups responsible for the challenges the authority faces in this area?

The structure described above is replicated to some degree at the local stations. It presents the intricate nature of the business of television, especially within the Nigerian context. It may differ in scale when there are no affiliated stations to consider, and the designated market area is much more compact, as is the case with the state government initiatives and some private concerns that emerged later. In the account presented above there has been little reference to written policies. Written policies make for smooth delegation of duty that becomes essential when organisations expand. Without clear channels of communication and stable policies, there will be distrust, unnecessary antagonism, backbiting and, in some quarters, apathy and discontent. These stifle creativity that the industry thrives on.

Though policies were written, in the NTA they were subject to interpretation, making it difficult to delegate duties. As a result, mentoring and staff progression were thorny areas. One would not expect the output of Programmes Directorate to be so sensitive that it accounts for a restriction in delegation of authority, yet it is. However, there is another reason offered by some. This is that delegation is hardly possible if there is inconsistency in treatment of staff and situations.

Q: Is that [the refusal of directors to delegate authority] on account of the sensitive nature of their office?

A: There are clear policies but because these are bent all the time, so [delegating officer] must see everything to know if the policy is to be bent [in that instance]. We tried to structure the Programmes Department in a modern way. You know, we tried to develop the Producers' Code of Conduct, [so that each person] —the VTR man, the Programmes Directors—each one was to have his own code. These things cannot be found easily in this organisation. I doubt

that it is on any file anywhere, except if individuals kept personal copies, and with this exodus of staff and general lack of contentment, where will you find such?"
(Excerpt from interview conducted in 1991)

In spite of the frustrations, some of the staff (including the above interviewee) kept faith with the organisation. That person is now a General Manager of a station. Such rich though challenging experiences may thus be drawn upon to improve structures. Therefore a charitable view of the situation over the years would regard the instability as a consequence of the growth in the organisation. Expansion on the scale witnessed in the organisation is bound to be accompanied by loss of instability in its processes. It is now time to consolidate.

Chapter 7

Postscript
Conclusions and Visions
for the Future

This book has presented, in some detail, the story of television in Nigeria. It shows that Africa deserves more than a passing mention when the medium is being discussed. Viewed against a backdrop of events on the political stage, the inextricable connection between this medium of mass communication and power becomes apparent. The book has shown that social transformations in different eras have not radically altered the mission of television. It has documented the joys and challenges of an industry. The story began with the observation that it was the political elite who recognised the potential of the medium and sought to harness its power at all costs. There is evidence of the growth in the industry and attempts to democratise access to what was once regarded as a medium for the privileged. This story has been told from an industry perspective, and some common themes have emerged in the different eras. These shall be highlighted in this chapter.

Though as a familiar medium, television needs no definition, this account of 50 years of television in Nigeria shows how complex the concept of television is. Those who encounter television in the 21st century may soon have faint, if any recognition of the type of experiences described here. For them, television is just another screen, like the computer screen. It may be synonymous with their games console; with time it may come to be likened to telephone screens. These are some of the new platforms on which television is used in contemporary times, even though some of the trends are yet to take root in Nigeria. With the sophistication in home-entertainment facilities, those who regard television as a form of mini-cinema can be excused. There is a blending of features in these days of media convergence, when television can be received on multiple platforms, each offering the viewer different controls, though experiences may be similar. For this reason, it is vital to be reacquainted with television. In this account is a reminder of how the medium was initially conceptualised— what was it meant to do.

Television is a medium of the 20th century. In 1959 it was novel, and though it was desirable even then, only a few could afford it, yet it was positioned as pivotal in the social development expected in post-colonial Nigeria. Television, like the other mass media, was to hasten the modernisation process in this context. As was the case in other parts of the world, the media were to replace (or simulate) traditional structures and channels of communication which had been displaced in the industrialisation process. This is the first evidence of some contradiction in the roles that television was meant to serve. It was meant to introduce new ways of life, some of which were seen to undermine traditional value systems. Yet it was also meant to draw legitimacy from these value systems to assert the cultural identity of the people. Indeed, finding resonance within the local cultures became essential for advocates of social change, a crucial aspect of television's role. Since the inception of television in Nigeria, the medium has bestridden two worlds—the local and global, the traditional and modern, the African and Western. Nigerian television thus reflects a hybrid of cultures even in the functions that it took on.

From media functions developed by Lasswell (1948) Wright (1960) and Mendelsohn, (1966) who identified the role of the media in (social) mobilisation, McQuail sets out some of the basic functions that the media play in contemporary society (McQuail 2006: 96 -8). In modern society, the media became the forum for receipt of information, the watchdog, the convener of different groups in society, the place to share common cultural experiences, determine distinct identities and examine values and the basis for permitting dissensions. The media were essential for maintaining social order, and they operated sometimes with groups or at the level of the individual. These were the functions performed by the church and the state in feudal Western society, but in the wake of industrialisation and the separation of church and state, there was a need for a neutral space in which these functions, particularly the rational discussion of ideas and opinion, could be conducted. The media need to be free to properly discharge their duties to society(Habermas 1989: McQuail 2006)

In the performance of their duties, the media became central in conferring status, fostering recognition for required roles and identifying talent and opportunities. By so doing the media facilitate the integration of different members or groups in society. The media also help to diffuse tensions evolving from the conflicts that are inevitable in the complex modern social systems that societies have become. The media's role is no less complex. In the same way that media identify heroes worthy of honour and credit, they also present villains who are deserving of condemnation. So the media inadvertently fuel resentment, distrust and rejection on the one hand, while promoting sympathy, acceptance and valorisation on the other. The media do not always set out in pursuit of such outcomes. Rather, they are deliberately deployed to provide information for specific social goals to be achieved; for example, when particular ways of life are promoted. The media tend to operate in the collective interest. This is the bedrock of public service broadcasting even when it has a commercial edge to it. Because broadcasting relies on public airwaves, a resource that belongs to none, its service must then be in the best interest of all. At other times broadcast media do serve private interest, in recognition of the sovereign rights of citizens. They could, for instance, offer individuals means of escaping from their reality, a chance to pursue their particular interests and dreams. This is to be expected in democratic societies where members can be regarded as citizens and consumers. However, what is by far the most important role of the media, particularly television, in a democratic society is its ability to convene an imagined community – people with shared experiences, shared memories and hopes (Turner, 2004). The structures around which programming is organised, whether they are local (regional), national or global, determine the scope of such communities. The Nigerian experience has shown that several factors are responsible for the structure of programming; some of these are the administrative units in the geographical area to be served, which could be economic (cost and source of programmes) and cultural, as is the case with the language of broadcast.

Indeed, the media are as complex as the society they serve, and they are imbued with the values of those contexts in which they are used. As Livingstone (1992) argued, media are socio-technical innovations, and their meaning must be drawn from the social dimension as well as the technical capabilities. This argument is similar to that advanced by those like Silverstone (1999) Morley (1986, 1992) and Moores (2000), who advocate an appreciation of media within the context of use—within domestic settings or routines of the ordinary. The same reasoning can be applied to an appreciation of the medium within public life and at national levels. Television does play a role in the construction of personal and national identities (Silverstone, Hirsch and Morley, 1992: 16). The differences in the contexts of its use thus offer useful clues in an appreciation of the mediating process and the significance of the different identities. This suggests that as universal as the technology of television may be, its meaning is likely to vary in different contexts.

Jefkins & Ugboajah (1986) offered some insight into the traditional media systems that operate in African contexts and frame media operations. These show structures that television producers have tried to simulate over the years. As discussed in Chapter 1, television was introduced as a teacher, an educator. In this role it usurped the position of the elders—sages who offered instruction to the children, guidance to the youth and leadership for all. It was also a forum for discussion and exchange of ideas, like the marketplace of old, the barbershop or drinking holes, any such place where people converge. Television was also to be the peddler of official information, the sort that was issued by the town crier of yore. It is a story-teller, too; its stories abound and they take various forms. It tells stories in the news; these include eyewitness accounts like those from itinerant traders, migrant fishermen and pastoralists, even long-distance lorry drivers. It presents a variety of experiences from near and far. Music, drama, dance and puppetry, used traditionally for entertainment and instruction, have been adopted for the medium. Traditional religious practices have survived on screen alongside the music and drama with which they were associated. The myths of religious festivals have been perpetuated through the publicity they have received on air, though it can be argued that televising some of these may undermine their mystical quality. However, all these cultural functions found expression in the new medium. In addition, television became a noticeboard for government officials and members of the community – citizens and guests alike. It was a place to feature their social engagements, chieftaincy titles and obituaries. Unlike newspapers, which required audiences to be literate, and radio, which presented disembodied voices, television was best placed to capture some of these experiences for audiences who, though resident in urban or semi -urban areas, had a rural mind set. In this was the justification for the visionary leaders who co-opted television for such lofty objectives; it is also its first triumph in the task set before it.

One point to be deduced from this excursion into television in an African context is the challenge of the cohabitation of different cultures. The challenge is one for producers as it is for audiences. With which culture should they pitch their loyalties? Once introduced from their Western origins, the media had to be aligned to address local audiences, perhaps in acknowledgement of the importance of the need to court audiences, taking cognisance of their preferences and cultural attitude to the use of time. After all, television viewing, if regarded as leisure, is subjected to cultural beliefs about time invested in leisure. Some cultural beliefs regard leisure of this sort as a waste of time—fancy sitting round a box for the express purpose of watching others engage in questionable activities. For some viewers, time invested in watching television must be justified by some recognisable benefit. Of course this orientation and their evaluating criteria may vary, depending on age, gender and other factors. These are elements in the matrix of considerations that informs scheduling and programming. However, at the initial stages there were limitations to what could be done about the flow of foreign programmes. From the outset, it seemed cheaper and perhaps more convenient for stations to be mere conduits of information and entertainment from abroad, even when the ideological implications of this were known. But the challenge of making programmes locally was faced squarely. This book offers glimpses into the effort and personal sacrifices that went into those productions. The evidence shows that television is a product of the many individuals who make up the production teams and labour to keep the stations on air. From the initial introduction of the medium right through the various stages of expansion, many have been charged with the responsibility of marshalling ideas for the medium. In this army, some were without the requisite training, yet they forged ahead because they were creative and talented. In time, amateurs and a dedicated corps of professionals who worked against enormous odds emerged in Nigeria. They were drawn from different backgrounds, different disciplines and different stations in life, yet they are judged by global standards, by its audiences and other observers. This is to be expected in an industry so integrated within global activities. A recurring concern even among television practitioners is the level of professionalism and training.

Demands for professional standards are easier to meet when certain aspects of the practice are codified. This arguably is what makes regulation of journalism and the more technical elements of television production less controversial, but much about television remains cultural. The discrepancy between the academy and the practitioners is partly due to the perception among certain elements within industry that some theories are impractical in the Nigerian context. Practitioners have been known to ridicule suggestions for shorter interview clips in their news bulletins. This suggestion was intended to enhance the quality of their news, with more concise reports. The practitioners were apprehensive of the breach, in their usual deference to interviewees (traditional rulers and other members of the elite) who might take offence if their speeches were edited too tightly (Bourgault, 1995: 50). There has been a shift since that observation was made. There are more private television stations in operation, which have demonstrated the viability of a different orientation. Even the Nigerian Television Authority (NTA), which attempts to report news from several bureaus around the nation, tends towards more reports and crisper sound-bites. With tele-prompters being used in more stations, there appears to be a greater attempt by newscasters to make eye contact with the audience, thus creating an impression that they are engaging in a dialogue. Casual observation of the news suggests that there are more stories with actualities—video reports of events covered—though the visual variety within individual packages remains doubtful in some instances. The use of eyewitness accounts and invited guests (analysts) on segments within newscasts has led to a blurring of the lines between news and current affairs programmes. This reorientation of the news may have been prompted by the increase in the length and number of newscasts in the light of longer transmission days. If this is what viewers want, then it must be regarded as progress.

The most striking evidence of change in television news in Nigeria is the shift from the deference to the elders and invited guests that characterised the media in an authoritarian regime. There is a tendency on some stations towards more provocative/adversarial interview styles. Depending on the topic, similar aggression is evident among passionate members of the public, who now have opportunities to participate on programmes in a variety of ways. Having emerged from the military dictatorship, Nigerians have learnt to have their say, if not their way. However, it is important that this liberty of expression be managed properly, in the interest of decorum.

Evaluations of television programmes that are based on more creative elements of production are subjective. It is more difficult to reach a consensus on these. There are bound to be complaints about television service based on different affinities among viewers. The jury is thus out on the achievements on other programmes. Opinions vary about television's performance, due to the large and diverse audiences, so Nigerian television gets a good bashing from those whom it seeks to serve. Government-owned stations appear to be worst hit because of their ties to government through funding, politicisation of management and allegations of sycophancy. This reputation is largely a product of the experience of managing stations under the military, which led to the overwhelming support for private broadcasting. Yet there is no guarantee that privately funded stations are immune to undue influence from political patrons, hence some of these stations also tend to be slated despite their effort. Between existing stations, audiences manage to get an array of political perspectives. This justifies deregulation and the pluralism that it has brought, particularly to more lucrative urban areas. The industry thus appears richer for the competition but there is another sort of competition that the nation needs to be wary of.

Access to satellite television stations (cable channels or bouquets on Digital Satellite Television, DStv) means that the more affluent audiences are exposed to foreign stations whose standards become the yardstick for Nigerian stations. This is evidence of one of the threats to the democratic potential of television imposed by global practices, as identified by Turner, who observed that "the role television can play within the imagined community of a nation-state has changed". (2004:5). The change is at the instance of a new menu of global programmes around which an imagined global community is being constructed, and these tend to undermine the distinctive efforts of national broadcasters. In this instance, South African-based MultiChoice Africa and its local partner in Nigeria are the most visible suppliers on the DStv platform. Though these companies have sympathetic objectives, and are supportive of independent producers, they inadvertently show up the shortcomings of the local stations on their bouquet. Through local content initiatives Multichoice is able to impart skills to local content creators.

The DStv offer includes premium channels such as BBC Worldwide, CNN, Discovery, Super Sports, Disney and Cartoon Network. MultiChoice also covers the UEFA Champions League, FA Cup, Carling Cup and English Premier League. How do local stations compete against these? Pay-TV and cable operators are able to benefit from the vertical integration that has characterised the media industry. These players are not encumbered by the same constraints that the more traditional public service-oriented stations had, for instance catering for all interests and tastes and ensuring that there is a balance in service, or even operating in recognition of the special duty that television stations have in constructing a sense of national identity and community. Unlike the generalist appeal of most Nigerian stations, their competitors on the new delivery platforms tend to be specialist channels that focus on particular genres—news, movies, sports, cartoons. Clearly the scales are tipped against terrestrial channels, which have contributed immensely to the development of local talent. They discover and groom talent, and continue to develop the local infrastructure in the arts.

Since the deregulation of broadcasting, more Nigerians have also taken up such opportunities afforded by the diversification of production. The National Broadcasting Commission licenses broadcasting operations at different levels in addition to that discussed in here. Hence, along with terrestrial stations, there are Nigerian-produced cable, satellite free-to-air television, terrestrial subscription television, satellite multi-channel subscription (direct-to-home) operations in Nigeria. These have not been explored fully in this account, and neither have the efforts made by foreign-based stations (such as Ben Television, Hi TV and Nollywood, all on BSkyB).

Television stations serve a wide variety of markets. This has been at great personal costs to producers throughout the years. The medium has a sterling record in programming and community service, won through teamwork, commitment and persistence as described in the following.

Here we work round the clock to ensure that the promoted programmes are transmitted. Even when we are one minute late, we must apologise. Come rain, come shine, we must be here. Even when somebody dies, we must just drag him aside and continue. If the transmitter packs up, engineers are called out in the middle of the night. The most annoying thing is that we don't have public holidays ... because even those whose sets were broken down ensure that they are fixed, so that they can enjoy good programmes over the period, even during weekends. There are fewer problems during the week, because people are too tired to be critical, but when people are relaxing at the weekends is the time we have to be on our toes. 'Igbadun ti yin, wahala ti wa' [meaning, Your pleasure is our pressure]. (Library Officer, Broadcasting Corporation of Oyo State interviewed in February 1991)

Audience adoption of television has been built on the labour creativity and shared commitment of a great many who have contributed to the development of the industry. Sadly, newcomers on the scene in every phase of its history build on opportunities created in the failings of previous generations of TV broadcasters, but even in this is evidence of growth. In 1959 and the early 1960s when the story of television began in Nigeria, there were great expectations from theorists like Schramm and Lerner that the mass media would foster development in less industrialised nations. The received wisdom in that era was rooted in the view that the media were all-powerful and that they were vital in the pursuit of modernisation and the acquisition of features of more industrialised nations. With the appropriate doses from the right media mix, societies could be nudged on to the level of social and economic growth desired. In time the deficiencies in such arguments were identified, as acquisition of mass media did not necessarily translate into the planned development goals. The failure of development interventions helped to revise theories on the nature and power of the media, since access to information alone was not sufficient in making a message effective. Realisation of such limitations helped to cast the beam on audiences and what they did with the media, what factors in their context had a bearing on planned communications. This has called attention to the context of reception; that is, what messages are required and how might these messages resonate with the audiences.

The Nigerian example has shown the relationship between the assumptions about the role that a medium would play and its rate of adoption. The initial push was made with the commitment of the political elite to using the television. This was supported to some degree by the faith of the business community in the benefits they could get from the venture; its opportunities in delivering desirable avenues of publicity accelerated the process. Support from the wider community, including cultural leaders, members of the academic community—those in the *avant-garde* for change—gave impetus to the diffusion of television. While the focus in this book had been on the effort to deliver the service, it is important to note that activities of these key facilitators of the innovation proved to be crucial in promoting acceptance of the service among audiences. Of course, any success in delivering service was only because audiences, both young and old, were fascinated by the prospects of television. For it was the initial curiosity about this magical box and subsequent subscription to its fare that kept the industry growing.

The box had both symbolic and functional value. Some people acquired it to enhance their status. Once government had given grants to help senior civil servants to boost TV set ownership at the initial stages of transmission in Nigeria, it was a matter of time before the affluent among the self-employed went on to acquire sets to prove their own financial status. Television became an essential item of furniture in the home. It is still sold on the sentiment of who has the better set ("My set is better than yours"), yet those for whom it was to be most beneficial were on the fringes. Television was adapted to be relevant to those privileged to own it. For viewers who were employed in public service—teachers, health workers, public administrators—the set had a high utility value, as it helped them to keep abreast of public information that the medium became known for. In a nation where government is the highest employer of labour and there is such extensive interdependence of the public and private sector, a substantial number of people would find such public information relevant. In any case, news remains an important aspect of television service. That was the traditional role of newspapers, which one can argue has been taken over by radio and television. Government patronage of television news thus gives it a sharper edge over the other media in this regard.

For many more people, television was an invaluable educational resource. This, after all, was the primary responsibility for which it was acquired. Television was a complement to schools, but it was also the school for life skills. It is a medium which audiences are prepared to learn from, regardless of the type of programmes they watch. People have been known to learn from drama, the news, discussion programmes, homilies and designated educational programmes. These focused on formal education, as well as practical lessons from agricultural extension workers. Then there were enlightenment programmes on a range of subjects such as health, nutrition, cultural practices, civic duties and legal rights. Television has also been known to be the place to hone creative skills. Budding artistes cut their teeth on the children's shows; adult performers have expanded their fan base through television. Many in the arts (musicians, performers, entertainers of different ilk), are indebted to the medium for the success and fame that they enjoy today. This was long before the reliance on reality TV shows to showcase talent and bring characters to the limelight.

In like manner, television has facilitated excursions to distant countries. It has played its part in creating the allure of foreign lands. This could be through documentaries, instructional television or even drama productions and films. The proportion of imported programmes was once so high that they attracted regulation. So grave was the concern about the flow of information that stations, with the NTA to the fore, set limits of their own volition. The quota as a rule was a ratio of 60% local content to 40% foreign programmes. Since its arrival, the National Broadcasting Commission has codified the minimum proportion for local content at 70%. The usual concern regarding the flow of information had been fear of cultural domination of the dependent industrialising nations. Little has been said about the repercussions of this flow on the dominant, more industrialised nations from which the programmes originate. For instance, are those societies more attractive destinations for migrants because media representations of them convey images of better living standards than what local societies can offer? This is uncharted territory in studies of the media and migration. Another area to be studied further is the penetration and impact of the emerging reverse flow of information from South to North. The international services from Nigeria exemplify this as they are well ahead of other African initiatives. Investment in international service from government and private broadcasters is justified by the magnitude of the human flight to these Western societies. At the moment Nigerian efforts are more realistically defined as service to diasporic audiences than a meaningful dialogue with other cultures. But the effort should not be despised, as a Pan-African audience base may be evolving gradually. The international service is in its infancy. Some of the issues involved are yet to be properly addressed, and they are beyond the scope of this book, but in these ventures Nigerians deserve financial and diplomatic support.

Much has been said about the technical dimensions to television service; at different levels the service had been compromised by the quality of available technology. The nation has enjoyed benefits of technological advances, in the use of Digital Satellite News Gatherers (DSNG), which facilitates live coverage, and the flexibility of initiating signals from multiple locations concurrently. This is one of the recent conquests of television in the new millennium. Before this, Nigerian television broadcasters had taken drastic steps to ensure survival, especially under the military. Many resorted to using home video camera systems. Though the quality was poor, they were able to maintain continuity in service in the days when they were starved of funds. There have also been accounts of the corruption that accompanied public spending in that era which meant equipment was purchased without spare parts. According to Bourgault, senior officers in broadcast organisations and the supervising ministries were often culpable in the situation that thereafter stifled creativity and frustrated the audience.

> Replacement of spare parts has often been a particular concern of management. One engineer of Harris Broadcasting of Illinois, a company supplying production and transmission equipment to the NTA, averred that spare parts and training for technical operatives

were the first components of telecommunication/broadcasting equipment packages to be bargained out of commercial agreements. Funds were very often siphoned away from these needs so as to make way for the 'gratuities' of well placed Information or Communication bureaucrats.

(Bourgault 1995: 52)

Clearly there are lessons to be learnt from this far-reaching revelation about past practices. It is illogical to procure equipment without adequate spares or training for the user departments and those technicians or engineers who maintain them. It is inconceivable that the same mistakes be repeated, given the length of experience in the industry. Where this happens, it undermines the culture of maintenance and morale in the organisation. It is those in the ranks who bear the burden of misplaced priorities.

When poor maintenance persists, staff have to rely on the sort of ingenuity that may have to go unnoticed. Because such procedures are unprofessional, they are simply not discussed. Such history should not be repeated if Nigeria is to maintain the leadership role that it has earned in the continent. It is ironic that while Nigeria jumped on the bandwagon of early adopters of television service and managed to keep up with global trends, as was the case with the introduction of colour transmission, it still struggles with technological issues. Its deadline to go digital has been set as 2012, though preparations for this goal are not as visible as they are in the UK, which also plans to switch off its analogue transmission in the same year. A number of stations have begun to invest in digital equipment and transmitters, but there is no evidence of the efforts regarding compatible and affordable television receivers. Some industry observers think digital broadcasting should not be a priority for Nigeria, given its other pressing needs— the economy, security and establishing a credible political culture. These sound like arguments that should have been advanced in 1959, when television service was launched, while most other African nations tarried a while. However, having got involved, there is little wriggle room out of global initiatives.

Costs involved in the business of television have escalated even further with the poor maintenance culture within television organisations and in the wider society. Where equipment is procured without necessary spare parts or training, as mentioned above, good ideas are snuffed out. No less crippling is the stifling of local initiatives by lack of funds or encouragement. However, the most notorious technical setback is one beyond the industry's control. This is the problem of poor electricity supply. Once a feature of less developed rural areas, "epileptic" power supply has become the bane of the nation's urban dwellers as well, but there is a bright side to this. With the influx of affordable petrol generators into the market, more people, including rural dwellers, can afford to generate electricity. Even the most lightweight generators afford their owners the opportunity to illuminate their room(s) and power refrigerators, electric fans and television sets. The value of this privilege is seen in the label given to them—*I better pass my neighbour* (Pidgin English, meaning, I am better than my neighbour). In this way, television has evolved from being the exclusive preserve of the privileged, as audience data show.

There is no denying that improved public infrastructure enhances media processes. The fact that stations must budget for private provision of what ought to be public infrastructure (road, electricity) is an unnecessary financial and logistical drain. Access to these has been known to enhance audience participation. Since improvements in Nigeria's telecommunications sector democratised access to mobile telephones and the internet, there has been an increase in the types of participation on television programmes. Whereas in the past audience participation was restricted to people who could physically be present in the studio or those whom the stations could reach, these days audiences can initiate contact and from the convenience of their location, they can phone in and send emails or SMS messages.

*Table 7.1: Average Time Spent Watching Television on Weekdays
By Age and Location January – March 2007*

Region	City	Under 18	18-24	25-40	Above 40	Total	Average
North East	Jalingo	183	124	133	161	601	150
South South	Calabar	0	154	117	107	378	95
South East	Owerri	130	121	116	111	478	120
North West	Zaria	180	133	137	121	571	143
North East	Gombe	210	143	155	140	648	162
North Cntrl	Ilorin	100	135	160	114	509	127
South South	Warri	101	111	110	171	493	123
South West	Abeokuta	120	96	110	123	449	112
South West	Akure	89	109	115	125	438	110
North West	Kebbi	150	153	150	131	584	146
Average		126	118	130	130	515	129

*Table 7.2: Average Time Spent Watching Television at Weekends
By Age and Location January – March 2007*

Region	City	Under 18	18-24	25-40	Above 40	Total	Average	Preferred Programme
North East	Jalingo	165	133	155	148	451	113	Superstory
South South	Calabar	0	177	126	121	424	106	Superstory
South East	Owerri	158	159	149	147	613	153	Superstory
North West	Zaria	180	162	188	190	720	180	Superstory
North East	Gombe	180	155	173	182	690	173	News
North Cntrl	Ilorin	147	72	162	182	563	141	Superstory
South South	Warri	121	150	147	139	557	139	News
South West	Abeokuta	152	126	144	82	504	126	Superstory
South West	Akure	127	125	140	144	536	134	Superstory
North West	Kebbi	130	170	159	153	612	153	Superstory
	Average	136	143	154	149	567	142	

Table 7.3: *Audience Preference - Best and Worst Programmes*
January – March 2007

Region	City	Preferred Programme	Worst Programme North
East	Jalingo	*Super Story**	Cartoons
South South	Calabar	*Super Story*	Adult Films/Movies
South East	Owerri	*Super Story*	Cartoons
North West	Zaria	*Super Story*	Political Discussions
North East	Gombe	News	Political Discussions
North Cntrl	Ilorin	*Super Story*	News
South South	Warri	News	Religious
South West	Abeokuta	*Super Story*	Cartoons
South West	Akure	*Super Story*	Political Discussion
North West	Kebbi	*Super Story*	Nigerian Films

*Super Story is a serial drama pitched as true-life matters of 'strife and sorrows'.
It is a Wale Adenuga Production a Lagos based independent production company but featured on NTA and other channels.
Data Courtesy National Broadcasting Commission Abuja Nigeria

In the final analysis, audience acceptance is the test of the merit of the service; that is the subject of another volume. However, it is gratifying to note indications that sections of the audiences respond positively, particularly in less populous (rural) towns. This book has presented an insight into various dimensions of television broadcasting. It has shown various forces in their struggle to take control of a medium that helps construct personal and group identity. It has shown that there is nothing intrinsically elitist about television, and its potential can truly benefit audiences at the grassroots in Africa. It has also highlighted simple lessons that should facilitate the progress and quality of service. For instance, whereas the issue of funding had been regarded as the obstacle to political independence required for public service television, the evidence suggests that with focus, commitment and prudent management, this need not be the case. With the willpower and a concerted effort, television can fulfill its myriad roles, among them promoting democracy and good governance. As the world moves up a notch in information and telecommunications, this reflection on the past should shed light on the visions for the future.

Bibliography

Abramson, A. (2002) *The History Of Television 1942 To 2000* Jefferson, NC: McFarland & Co.

Adebimpe, D. (2006) Commercialisation and Television: The OGTV Experience in Biodun Odetoyinbo (ed) *GTV: A Gateway to Broadcast Excellence* Abeokuta: GTV, 235 – 259.

Ajayi, J. F. A. (2000) Ethnicity and Nationalism in Nigeria in Toyin Falola (ed) *Tradition and Change in Africa The Essays of J F A Ajayi* Trenton: Africa World Press, 259 – 276.

Akinfeleye, R. (1999) Emerging Issues and Neglected Areas in Television for Development in Nigeria paper delivered at Forty Years of Television in Nigeria symposium organised by the Obafemi Awolowo Foundation.

Akinfeleye, R. (1996) Journalism Education and Training in Nigeria: Infrastructures, Policies and Development in Olatunji Dare and Adidi Uyo (eds) *Journalism in Nigeria: Issues and Perspectives* Lagos: Nigeria Union of Journalist, 228 - 267.

Amana, E. in interview with Jenson Okereke & Grace Rwang Grounded Transmitters: Amana gives insight NTA Corporate Newsletter July 2007 front page.

Aris, A. & Bughin, J. (2009) *Managing Media Companies; Harnessing Creative Value* Chichester John Wiley.

Atoyebi, B. (2002) State of Broadcasting in Nigeria in *Broadcast Regulation in Nigeria* Abuja: National Broadcasting Commission, 4 – 23.

Atte, J. & Emakpore M. (eds) (2006) *Political Broadcast Manual* Abuja: Nigeria Television Authority.

BBC World Service Website http://www.bbc.co.uk/worldservice/faq/news/story/2005/08/050810_wsstart.shtml.

Berenger, R. D. & Labidi K. (2005) Egypt in Anne Cooper (eds) *Global Entertainment Media Content Audiences & Issues* London: Lawrence Erlbaum, 81 – 98.

Block, P. (ed) with Houseley W. Nicholls, T & Southwell, R. (2001) *Managing in the Media* Oxford: Focal Press.

Borisade, O. (2006) Television Medium Management in Biodun Odetoyinbo (ed) *GTV: A Gateway to Broadcast Excellence* Abeokuta: GTV, 137 - 162.

Bourgault, L. M. (1995) *Mass Media in Sub-Saharan Africa* Bloomington Indiana University Press.

Boyd, A. (2001) *Broadcast Journalism Techniques of Radio and Television News* Oxford Focal Press 5th Edition.

Brann, C. M. B. (1985) A Sociolinguistic Typology of Language Contact in Nigeria: The Role of Translation in Frank Ugboajah (ed) *Mass Communication, Culture and Society in West Africa*, Oxford: Hans Zell Publishers,122 – 132.

Briggs, A. (1995) *The History Of Broadcasting In The United Kingdom* Oxford: Oxford University Press.

British Film Institute Data base http://ftvdb.bfi.org.uk/sift/organisation/ 37936.

Chu, G. C. and Schramm, W (1991) *Social Impact of Satellite Television in Rural Indonesia* Singapore Asian Mass Communication Research and Information Centre.

Chu, G. C. (1994) Communication and Development: Some Emerging Theoretical Perspectives in Andrew A. Moemeka (ed) *Communication for Development: A New Pan Disciplinary Perspective* Albany: State University of New York Press, 34 – 53.

Corner J. (1991) Television & British Society in the 50s in John Corner (ed) *Popular Television in Britain* London: British Film Institute, 1 -21.

Count Down WHO Newsletter on Polio Eradication in Nigeria No. 6 December 2008.

Cowling, L. (2005) South Africa in Anne Cooper (eds) *Global Entertainment Media Content Audiences & Issues* London: Lawrence Erlbaum Associates Publishers, 115 – 130.

Crissel, A. (2002) *An Introductory History Of British Broadcasting* (2nd edn) London: Routledge.

Curran, J. & Seaton J (2003) *Power Without Responsibility: The Press Broadcasting & New Media In Britain* (6th edn.) London: Routledge.

Curran, J. (1993) Rethinking the Media as a Public Sphere in P Dahlgren and C Sparks (eds) *Communication and Citizenship* London: Routledge, 27 – 57.

Curran, J. (1996) Mass Media and Democracy Revisited in J. Curran and M Gurevitch (eds) *Mass Media and Society* (2nd edn.) London: Edward Arnold, 81 – 119.

DAAR Communications PLC Prospectus for Initial Public Offering February 2008.

Dahlgren, P. (1995) *Television & the Public Sphere: Citizenship, Democracy and the Media.*

Dahlgren, P. (2000) Key Trends in European Television in Jan Wieten, Graham Murdock and Peter Dahlgren (eds) *Television Across Europe* London: Sage, 23 – 34.

Dahlgren, P. (2002) The Public Sphere as Historical Narrative in Denis McQuail (ed) *McQuail's Reader in Mass Communication Theory* London: Sage, 195 - 200.

Dahlgren, P. (1995) *Television and the Public Sphere* London: Sage.

De Goshie, J. (1986) *The Use of Radio and Television as Educational Materials: The Case of National Educational Technology Centre (NETC) Kaduna* Unpublished paper presented at the AERLS Agricultural Information Workshop Ahmadu Bello University.

Didigu, F. (2002) Broadcasting and Family Values in Nigeria in *Broadcast Regulation in Nigeria* Abuja: National Broadcasting Commission, 106 - 133.

Dominick, J. (1996) *The Dynamics of Mass Communication* New York: McGraw Hill Inc 5th Edition.

Ebuetse, C. (2007) Raymond Aleogho Dokpesi: *A Legend of Our Time* Ibadan: Spectrum Books.

Egyptian State Information Service Radio and Television http://www.sis.gov.eg/En/Arts&Culture/RTV/071700000000000001.htm (last accessed 15th April 2009).

Eldridge, J. Kitzinger, J. & Williams K. (1999) *The Mass Media and Power in Modern Britain* Oxford: Oxford University Press.

Eleshin, K. (2006) in Biodun Odetoyinbo (ed) *GTV: A Gateway to Broadcast Excellence* Abeokuta: GTV, 122 – 136.

Enemaku, S. (2006) Television an Instrument for Development in Biodun Odetoyinbo (ed) *GTV: A Gateway to Broadcast Excellence* Abeokuta: GTV, 43 - 60.

Esan, O. (1993) *Receiving Television Messages: An Ethnographic Study of Women in a Nigerian Context* (Unpublished Doctoral Thesis) University of Glasgow.

Fadakinte, M. M. (2002) The Nigerian State and Transition Politics in Browne Onuoha & M.M. Fadakinte (eds) *Transition Politics in Nigeria 1970 – 1999* Lagos: Malthouse Press, 40 – 60.

Fakiyesi, T. (2002) Nigerian Economy and Political Transition in Browne Onuoha & M.M. Fadakinte (eds) *Transition Politics in Nigeria 1970 – 1999* Lagos: Malthouse Press, 219 – 231.

Falola, T. and Heaton, M. (2008) *A History of Nigeria.* Cambridge: Cambidge University Press.

Fiske, J. (1987) *Television Culture* London: Routledge.

French, D & Richards M (1994) Theory Practice and Media Forces in Britain: A Case of Relative Autonomy in D French and M. Richards (eds) *Media Education Across Europe* London: Routledge, 82 – 102.

Habermas, J. (1989) *The Structural Transformation of the Public Sphere* Cambridge: Polity.

Hussein, A. Egypt on The Museum of Broadcasting Communications http://www.museum.tv/archives/etv/E/htmlE/egypt/egypt.htm (last accessed 15th April 2009).

Iredia, T. (2004) *Public Broadcasting in a Developing Nation A Focus on the Nigerian Television Authority* public lecture given at the September 2004 Convocation Ceremony of the Benson Idahosa University Benin City.

Jakande, L. (2007) The Travails of LTV in *LTV Next* April 2007 :6

Jefkins, F & Ugboajah F (1986) *Communication in Industrialising Countries* London: Macmillan.

Kolade, C. (1985) in TJ Interview with Dr. C. Kolade *Television Journal* (July / September 1985) [Nigerian Television Authority].

Larkin, B. (2008) *Signal and Noise: Media, Infrastructure and Urban Culture in Nigeria* Durham: Duke University Press.

Lewis, D. (2006) I was used and dumped – Dejumo Lewis Oloja, Village Headmaster interview with Onyekaba Cornel-Best Daily Sun (Nigeria) Friday 17 2007 http://www.sunnewsonline.com/webpages/features/showtime/2006/feb/17/showtime-17-02-2006-001.htm (last accessed 12th August 2009).

Livingstone, S. (1992) The Meaning of Domestic Technologies, in Roger Silverstone & Eric Hirsch (eds) *Consuming Technologies: Media and Information in Domestic Spaces* London: Routledge, 113 – 130.

Maduka, V. (1999) Forty Years of Television in Nigeria: The Promise and Performance Keynote paper at symposium organised by the Obafemi Awolowo Foundation.

Manning, P. (2001) *News and News Sources.* London: Sage.

McMurria, J. (2004) Global Channels in J. Sinclair & G. Turner (eds) *Contemporary World Television* London: British Film Institute, 38 – 41.

McQuail, D. (2002) *McQuail's Reader in Mass Communication Theory* London: Sage, 3 -18.

McQuail, D. (2005) *Theories of Mass Communication* London: Sage.

Moemeka, A. (1994) Development Communication: A Historical and Conceptual Overview in Andrew A. Moemeka (ed) *Communication for Development: A New Pan Disciplinary Perspective* Albany: State University of New York Press, 3 – 22.

Moemeka, A. (1994) Development Communication: Basic Approaches and Planning Strategies in Andrew A. Moemeka (ed) *Communication for Development: A New Pan Disciplinary Perspective* Albany: State University of New York Press, 54 – 73.

Moores, S. (1997) Broadcasting and its audiences in H McKay (ed) Consumption and Everyday life London: Sage, 213 – 246.

Moores, S. (2000) *Media and Everyday Life in Modern Society* Edinburgh University Press.

Morley, D. (1986) *Family Television: Cultural Power and Domestic Leisure* London: Comedia.

Morley, D. (1991) Changing Paradigms in Audience Studies in Ellen Seiter, Hans Borchers, Gabriele Kreutzner, Ana-Maria Warth (eds) *Remote Control Television: Audiences and Cultural Power* London: Routledge, 16 – 41.

Morley, D. (1992) *Television Audiences and Cultural Studies* London: Routledge.

Mukoro, T (2008) How I became the First Village Headmaster Interview with Nigeria Films.com in ModernGhana.com Monday May 26 2008 http://www.modernghana.com/movie/2373/3/how-i-became-the-first-village-headmaster-ted-muko.html (last accessed 12th August 2009).

Multichoice Africa Stakeholders Report.

MultiChoice Nigeria (2008) Delivering quality Pay Television in Nigeria MultiChoice Nigeria Stakeholders Report.

Murdock, G. (2000) Talk Shows, Democratic Debates and Tabloid Tales in Jan Wieten, Graham Murdock and Peter Dahlgren (eds) *Television Across Europe* London: Sage, 198 – 220.

Mytton, G. (1983) *Mass Communication in Africa* London: Arnold.

Mytton, G., Teer-Tomaselli, R and Tudesq A-J (2005) Transnational Television in sub-Saharan Africa in Jean K Chalaby (ed). *TransNational Television Worldwide: Towards a New Media Order* London: IB Taurus, 96 – 127.

National Population Commision (NPC) [Nigeria] and ORC Macro 2004 *Nigeria Demographic and Health Survey 2003* Calverton, Maryland: National population Commisssion and ORC Macro.

Nigeria direct website http://nigeria.gov.ng/NR/exeres/D53A68AB-3C7C-4B2D-934E-3762891DC995.htm.

Nigerian Broadcasting Code 2004 (4th edn) Abuja: National Broadcasting Commision.

Nordenstrong, K. & Varis, T. (1974) *Television Traffic - A One-Way Street?* Paris: Unesco.

NTV Ibadan (1979) Obaro Ikhime (ed) *20th Anniversary History of WNTV* Ibadan: Heinemann.

Obazele, P. (1996) Challenges of Radio Journalism and Management of Broadcasting in Nigeria in Olatunji Dare and Adidi Uyo (eds) *Journalism in Nigeria: Issues and Perspectives* Lagos: Nigerian Union of Journalists, 144 – 158.

Okereke, J (2007) NTA Monitoring Team Swings into Action in NTA Corporate Newsletter Vol. 1 No.2 July 2007.

Olasope, K. (2009) *Africa TV at 50 Project* Nigerian Tribune 9th of June 2009 http://www.tribune.com.ng/09062009/opinion2.html (last accessed 12th August 2009).

Olorunnisola, A. A, & Akanni T. M. (2005) Nigeria in Anne Cooper (eds)

Global Entertainment Media Content Audiences & Issues London: Lawrence Erlbaum, 99 – 114.

Olukoju, A. (2004) Nigerian Cities in Historical Perspectives in Toyin Falola & Steven J. Salm (eds) *Nigerian Cities* Trenton NJ: Africa World Press, 11 – 46.

Onabanjo, O. (2006) Television is for Public Interest and Good in Biodun Odetoyinbo (ed) *GTV: A Gateway to Broadcast Excellence* Abeokuta: GTV, 1 – 14.

Oni, D. (1985) The Bala Miller Show in *Television Journal* No.8 April – June 1985 14 – 35.

Onuoha, B. (2002) Reflections on the Transition Programmes in Browne Onuoha & M.M. Fadakinte (eds) *Transition Politics in Nigeria 1970 – 1999* Lagos: Malthouse Press, 19 – 39.

Oso, L. (2006) Between Commercialisation and Social Responsibility: Finding an Equilibrium in Biodun Odetoyinbo (ed) *GTV: A Gateway to Broadcast Excellence* Abeokuta: GTV, 261 – 294 .

Petley, J. (2006) Public Service Broadcasting in the UK in Douglas Gomery & Luke Hockley (eds) *Television Industries* London: British Film Institute , 42 – 45.

Preston, A. (2004) Evolution Note Revolution: The Ecology of Multichannel Television in J. Sinclair & G. Turner (eds) *Contemporary World Television* London: British Film Institute, 45 – 49.

Price, E. M. (1995) *Television, the Public Sphere and National Identity* Oxford: Oxford University Press.

PRTV (2005) Mannok J et al. (eds) *Media: A Passion for Peace a Commitment to Serve* Jos: St Stephen Bookhouse Inc.

Rohdes & Schwarz website Rohdes and Schwarz http://www2.rohde-schwarz.com/ (last accessed 15th April 2009).

Ronning, Helge & Kupe, Tawana (1999) The Dual Legacy of Democracy and Authoritarianism in Myung Jin Park & James Curran (eds) *DeWesternizing Media Theory* London: Routledge, 157 – 177.

Saidu, R. (2002) The NBC Act, Code and Legal Challenges in *Broadcast Regulation in Nigeria* Abuja: National Broadcasting Commission, 24 – 50.

Salama, G. (1978) *Television in a Developing Society* Jos: Nigerian Television Authority.

Salihu, L. (2002) The Impact of Foreign Broadcast Media on Broadcasting in Nigeria in *Broadcast Regulation in Nigeria* Abuja: National Broadcasting Commission, 159 - 172.

Salihu, M. A. (2002) Broadcasting and Crisis Management in *Broadcast Regulation in Nigeria* Abuja: National Broadcasting Commission, 134 – 158.

Scannell P. & Cardiff D. (1991) A Social History of British Broadcasting Oxford: B. Blackwell.

Seaton, J. (2003) Broadcasting and the Theory of Public Service in James Curran & Jean Seaton (eds) *Power without Responsibility* London: Routledge, 363 – 376.

Siebert, F., Peterson, T. and Schramm, W. (1956) *Four Theories of the Press* Urbana, Il: University of Illinois Press.

Silverstone, R. (1999) *Television and Everyday Life* 2nd edition London: Routledge.

Silverstone, R., Hirsch, E. & Morley, D. (1992) Information and Communication Technologies and the Moral Economy of the Household in Roger Silverstone & Eric Hirsch (eds) *Consuming Technologies: Media and Information in Domestic Spaces* London: Routledge, 15 – 31.

Simpson, E. (1985) Translating in Nigerian Mass Media in Frank Ugboajah (ed) *Mass Communication, Culture and Society in West Africa*, Oxford: Hans Zell Publishers, 133 – 152.

Sotumbi, B. (1999) The Content and Quality of Television Programmes in Nigeria peper delivered at Forty Years of Television in Nigeria symposium organised by the Obafemi Awolowo Foundation.

Sotumbi, B. (1996) The Challenges of Television Journalism in Olatunji Dare |& Adidi Uyo (eds) *Journalism in Nigeria: Issues and Perspectives* Lagos:Nigeria Union of Journalists, 172 – 183.

Sparks, C. (1999) Media Theory After the Fall of Communism in Myung Jin Park & James Curran (eds) *DeWesternizing Media Theory* London: Routledge, 35 – 49.

Television Journal July / September 1985 Interview with Christopher Kolade.

Thussu, D. K. (2004) Television and Local Imagined Communities in J. Sinclair & G. Turner (eds) *Contemporary World Television* London: British Film Institute, 28 – 31.

Thussu, D.K. (1998) Infotainment International: A View from the South in D. K. Thussu (ed) *Electronic Empires* London: Arnold, 63 – 82.

Tomaselli, K.G & Heuva, W (2004) Television in Africa in J. Sinclair & G. Turner (eds) *Contemporary World Television* London: British Film Institute, 96 – 99.

Uche, L. (1998) *Mass Media People and Politics in Nigeria* New Delhi Concept Publishing.

Ugboajah, F. O. (1985) Oramedia in Africa in Frank Ugboajah (ed) *Mass Communication, Culture and Society in West Africa*, Oxford: Hans Zell Publishers, 165 – 176.

Ugboajah, F. O. (1980) *Communication Policies in Nigeria* Paris: United Nations Educational Scientific and Cultural Organisation.

Ume-Nwagbo, E. N. E. (1986) *"Cock Crow at Dawn" A Nigerian Experiment With Television Drama in Development-Communication* in International Communication Gazette Vol. 37 No 3 (1986) 155 – 167.

Uzodinma, I. (ed) (1981) *NTA Handbook 1981* Lagos Nigerian Television Authority Division Corporate Affairs.

Western/Nigeria Radiovision Services *Annual Reports and Accounts 1968/69* of Western Nigeria Government Broadcasting Corporation.

Williams, R. (1990) *Television Technology and Cultural Form* London: Routledge.

Winston, B. (2002) Towards Tabloidization? Glasgow Revisited 1975 – 2001 *Journalism Studies* Vol. 3 No.1: 5 – 20.

WNBS/ WNTV Accounts 1972 Ibadan Western Nigeria Radiovision Services Ltd.

Wright, C. (2002) The Mass Society in McQuail D (ed) *McQuail's Reader in Mass Communication Theory* London Sage 73 – 79.

Index

Lightning Source UK Ltd.
Milton Keynes UK
UKOW021456190112

185697UK00011B/52/P